Eating Rice from Bamboo Roots

The Social History of a Community

of Handicraft Papermakers in

Rural Sichuan, 1920–2000

Harvard East Asian

Monographs 314

Eating Rice from Bamboo Roots

The Social History of a Community of Handicraft Papermakers in Rural Sichuan, 1920–2000

Jacob Eyferth

Published by the Harvard University Asia Center
Distributed by Harvard University Press
Cambridge (Massachusetts) and London 2009

Printed in the United States of America

The Harvard University Asia Center publishes a monograph series and, in coordination with the Fairbank Center for Chinese Studies, the Korea Institute, the Reischauer Institute of Japanese Studies, and other faculties and institutes, administers research projects designed to further scholarly understanding of China, Japan, Vietnam, Korea, and other Asian countries. The Center also sponsors projects addressing multidisciplinary and regional issues in Asia.

Library of Congress Cataloging-in-Publication Data

Eyferth, Jan Jacob Karl, 1962–
    Eating rice from bamboo roots : the social history of a community of handicraft papermakers in rural Sichuan, 1920–2000 / Jacob Eyferth.
        p. cm. -- (Harvard East Asian monographs ; 314)
    Includes bibliographical references and index.
    ISBN 978-0-674-03288-0 (cl : alk. paper)
    1. Paper industry--China--History. 2. Paper industry--China--Jiajiang Xian--History. 3. Jiajiang Xian (China)--Economic conditions.
    I. Title.
        HD9836.C62E94 2009
        338.7'6760951380904--dc22
                                                    2009007044

Index by Mary Mortensen

♾ Printed on acid-free paper

Last figure below indicates year of this printing
19 18 17 16 15 14 13 12 11 10 09

For my parents

Acknowledgments

FOR PAPERMAKERS IN PRE-1949 Jiajiang, debt was not necessarily a bad thing. To be in debt was to be in business, to participate in long-term relations of exchange. I, too, am happy to acknowledge my debts, even though I realize that many of them will never be repaid. My greatest debt is to the people of Shiyan who welcomed me into their homes, answered endless questions, and in the process taught me much more than I realized at that time. Much as I would like to extend individual thanks to them, I decided to preserve the anonymity of all interviewees. Such reserve, I felt, is not necessary in the case of former officials in the county seat, many of whom are local historians in their own right. Ren Zhijun, Xiao Zhicheng, Xie Baoqing, Xu Shiqing, Huang Fuwan, and Liao Tailing of the Second Light Industry administration shared their immense knowledge of the paper industry with me. Zhang Wenhua and Xie Changfu, the compilers of the township gazetteers of Macun and Huatou, gave me lively lessons in local history. Fieldwork would have been impossible without the support of the village committee of Shiyan, the Macun township government, and the Foreign Affairs Bureau of Jiajiang. I also want to thank my hosts at the Institute of Rural Economy of the Sichuan Academy of Social Sciences (SASS) for arranging fieldwork at a time when research permits were hard to obtain. Du Shouhu, Guo Xiaoming, and Zhang Xiangrong repeatedly traveled to Jiajiang to sort out administrative problems and gave me much practical advice. Heartfelt thanks are due to the SASS researchers who accompanied me to Jiajiang: Chen Jining, who gave me invaluable lessons in rural etiquette, and Lei Xiaoming, who spent months away from his family and fell ill during our stay but never complained. Back in Chengdu, Professors Wang Gang, Sheng Yi, and Yuan Dingji at the SASS and Ran Guangrong, Zhang Xuejun, and Wang Yan at Sichuan University helped me to locate sources and provided much-needed background information.

This book began its life as a dissertation guided by Tony Saich and Frank Pieke, both of whom provided guidance, encouragement, practical help, and hospitality in Oxford, Boston, and Beijing. Ward Vermeer, Tak-Wing Ngo, Barend ter Haar, Woei-Lien Chong, and Axel Schneider read all or parts of the dissertation and gave valuable comments. Equally important were the discussions with my fellow students at Leiden University, especially Hein Mallee, Peter Ho, Wu Yongping, and Yuan Bingling. Only now do I realize how much I benefited from the financial, institutional, and intellectual support of the (now sadly defunct) Centre for Non-Western Studies in Leiden. Many thanks to its director Willem Vogelsang and the CNWS staff for giving me almost unlimited freedom to write and travel for five years.

Since I left Leiden, I received generous support from the Institute for Chinese Studies of the University of Oxford, the Fairbank Center for Chinese Studies at Harvard University, and the Center for Historical Analysis at Rutgers University. Heartfelt thanks to the staff of these institutions and to their directors Glen Dudbridge, Liz Perry, Susan Schrepfer, and Phil Scranton. Research would have been impossible without access to the libraries of Leiden University, the Bodleian Library, the Fairbank Research Library, the Harvard-Yenching Library, and the libraries of Cornell University, Simon Fraser University, and the University of British Columbia. In China, I used the Sichuan Provincial Library, Sichuan Provincial Archives, Chinese National Library, and the Library of Sichuan University. The help of the staff of these libraries was invaluable and is gratefully acknowledged.

For comments, criticism, and encouragement, I thank Rana Mitter, David Faure, and Henrietta Harrison at Oxford; my fellow An Wang fellows Wang Liping, Zwia Lipkin, Sabina Knight, and Elizabeth Remick at the Fairbank Center; and Kavita Philip and Erin

Clune at the Center for Historical Analysis at Rutgers. I am indebted in so many ways to friends and colleagues at the history department of Simon Fraser University that I hardly know where to begin. As department chair, John Craig gave me research time when I most needed it. Felicitas Becker, Elise Chenier, Luke Clossey, Alec Dawson, Karen Ferguson, Andrea Geiger, Mary Ellen Kelm, Thomas Kühn, Mark Leier, Jack Little, Janice Matsumura, Sheilagh MacDonald, Emily O'Brien, Roxanne Panchasi, Helena Pohland-McCormick, Heather Skibeneckyj, John Stubbs, and Ilya Vinkovetsky all provided insights and encouragement. At the University of British Columbia, Ken Foster, Amy Hanser, Tim Cheek, Tim Brook, Tim Sedo, and the other participants of the China Study Group gave me critical feedback at a crucial time.

Over the years, I presented versions of parts of the book in many different contexts. Stig Thøgersen at the University of Århus and Claude Aubert at the Institut national de la recherche agronomique in Paris gave me first opportunities to try out ideas in public. Susanne Brandstaetter at the University of Manchester, Dorothy Ko and Eugenia Lean at Columbia, Fang Lili at the Chinese Art Academy, Ching Kwan Lee at the University of Michigan, Vivienne Shue at Cornell University, Wang Yuning at the Shaanxi Art Academy, and Diana Lary at the University of British Columbia allowed me to present my research in workshops, seminars, and panels they organized. I particularly thank Dorothy Ko for inspiration and encouragement, and Eugenia Lean for allowing me to quote from her unpublished work. I also benefited in many different ways from discussions with Maria Edin, Mareile Flitsch, Lida Junghans, Mark Frazier, Greg Ruf, Bin Wu, and Janet Sturgeon. Special thanks are due to Mark Selden, who read an early incarnation of the book with lightning speed, sent me detailed and insightful comments, and has provided feedback and encouragement at several points over the years.

Writing, like papermaking, is often a family enterprise. Paola Iovene has seen this book develop almost from the beginning and has contributed to it in more ways than she thinks. Much of the final writing took place on a terrace overlooking the Mediterranean Sea, thanks to the generous hospitality of Teresa Iacono and Giuseppe Iovene. Tobias Eyferth read and commented on different versions of the manuscript; Konrad Eyferth drew the maps. This book could not have been written without the support of my parents, Ina and Klaus Eyferth. I dedicate it to them.

J.E.

Contents

Reference Matter

Tables, Maps, and Illustrations

Weights, Measurements, and Money

BEFORE THE INTRODUCTION of the metric system in the late 1930s, weights and measurements in Sichuan differed not only between localities but also between trades, with substantial differences in the length of a carpenters' and a tailors' *chi*, or the weight of a *jin* of oil and a *jin* of rice. See Jerome Ch'en, *The Highlanders of Central China*, for a full discussion of weights, measures, and monies in Sichuan.

Grain volumes: 1 *shi* (locally pronounced *dan*, often translated as "bushel") equaled 10 *dou* (peck) or 100 *sheng* (pint). A Jiajiang *shi* of husked rice weighed approximately 90 kg.

Weight: 1 *dan* (often translated as "picul") was the load an adult man would carry with a shoulder pole, circa 60 kg. It was subdivided into 100 *jin* (pound) of 16 *liang* (ounce; also translated as "tael") each.

Length: One *li* was about one-half of a kilometer or one-third of a mile. One *chi* (foot) ranged from 30 to 34 cm.

Area: One *mu* was 0.15 acres or 0.066 hectare, subdivided into 10 *fen*. Land in Jiajiang was also measured in *dou* (pecks of seed grain). One *dou* of rice land was about 1.2 *mu*.

Money: Uncoined silver was measured in *liang* of 37.8 grams. Various silver coins modeled on the Mexican silver dollar were in use. Known as silver *yuan* or *dayang*, they weighed approximately 24 grams of silver.

Paper: Writing and printing paper was sold in *dao* of 100 sheets. Ritual paper money was sold by the *tiao* (load; 35–40 kg) or the *wan* (10,000—conventionally understood to be 9,000 or 9,500 sheets, depending on the type of paper).

Eating Rice from Bamboo Roots

The Social History of a Community

of Handicraft Papermakers in

Rural Sichuan, 1920–2000

Introduction

THIS BOOK CHARTS THE twentieth-century history of a community of rural artisans in Jiajiang county, halfway between Chengdu and Leshan in China's Sichuan province. At its center is the slow and painful work of men and women who transform bamboo stalks and other fibrous matter into soft, supple sheets of paper. Papermaking is highly skilled work, and the theme of skill runs through this book in two connected strands. I am interested, first of all, in production-related skills, which may be technical (how to mix pulp, how to mold a sheet of paper) or social (how to find buyers for your product, how to get along with your neighbors). Beyond that, I am interested in what could be called the skills of everyday life: the quotidian strategies that have allowed Jiajiang papermakers to survive and sometimes prosper despite war, revolution, and extraordinarily rapid social and economic change. These types of skills are linked, and a focus on the concrete details of skilled work gives us insights into the lifeworlds of rural people, whose experiences might otherwise remain concealed.*

Although I focus on the material and the everyday in one concrete place and situate the book in the rich tradition of Chinese village studies,[1] I am pursuing a larger argument. Throughout the book, I argue that the Chinese revolution—understood as a series of interconnected political, social, and technological transformations—was

---

*I provisionally define skill as "practical knowledge" or "knowledgeable practice," the kind of tacit, subjective, context-dependent knowledge that guides practitioners in their daily activities. Such skills may or may not be production-related. See Ingold, *Perception of the Environment*, 316, 352–54. See also Chapter 1 for a more extensive discussion of skills in the paper industry.

as much about the redistribution of skill, knowledge, and technical control as it was about the redistribution of land and political power, and that struggles over skill in twentieth-century China resulted in a massive transfer of technical control from the villages to the cities, from primary producers to managerial elites, and from women to men.

The largest context for this study is what is known among students of modern China as the rural-urban divide: an institutional, social, and economic cleavage that separates China's rural people (including the millions who have moved to the cities but remain linked, through the household registration system, to their rural homes) from China's urbanites.[2] This gap is as large as that dividing Chinese urban residents from residents of urban centers in the West, and despite recent changes in the institutions that created it, it shows no sign of closing. My argument in this book is that this rural-urban gap is caused in part by changes in the distribution of knowledge between rural and urban China that date to the beginning of the twentieth century and intensified after the socialist revolution of 1949. China, most historians now agree, underwent a long process of proto-industrial development not unlike that documented for Western Europe and Japan.[3] In Qing (1644–1911) and Republican (1911–49) China, as in Tokugawa Japan and in Europe before the nineteenth century, most manufactured goods originated in the countryside, in the households of farmers or semi-specialized farmer-artisans. In contrast to Western Europe and Japan, where handmade goods had largely been replaced by factory products by the end of the nineteenth century, Chinese craft industries survived relatively intact into the middle of the twentieth century. Maoist rhetoric holds that China's traditional industries collapsed under the onslaught of cheap foreign goods after the Opium War, but the available evidence suggests that "in absolute terms, handicraft output as a whole held its own or even increased" in the Republican period, though its relative share in the economy declined as China developed a modern industrial sector.[4] In 1933—the last "normal" year before China was hit by depression and war—handicrafts still accounted for three-fourths of industrial output.[5] Even as late as 1952, when China's modern industrial sector had begun to recover from the effects of the war and the revolution, handicraft industries accounted for 42 percent of industrial output in current prices (heavily weighted in favor of heavy industry) and for an amazing 68 percent in prewar prices.[6]

Despite—or perhaps because of—their economic staying power, rural industries came to be seen as deeply problematic by the Western-educated elites who dominated China after 1900. Humiliated by Japan's 1895 defeat of China and inspired by the example of the West and Japan, Chinese elites began to think of "the economy"—which in these years emerged as a distinct category separate from the social, cultural, and moral—as a staging ground for competition between nations.[7] Looking toward Europe and Japan, they came to view national economies as composed of sharply delineated yet complementary sectors. Industry was the leading sector, since it alone propelled nations toward a better future; it was typically urban, at home not in peasant households but in large, mechanized factories. The countryside, by contrast, was the realm of farmers who produced food for the nation but could not or should not produce industrial goods to any significant degree.[8] This view of an economy with distinct rural and urban sectors was inaccurate as a description of Chinese realities, but it was a powerful prescription for change. It is perhaps best seen, following James Scott, as a "state simplification": modernizing states, Scott argues, tend to translate complex social facts into simplified representations—maps, statistics, population registers—that make society "legible" and thus easy to control. To some extent, such simplifications are necessary, but problems arise when the state confuses its abstractions with the facts on the ground, or even sees them as some higher form of order to which observed realities must conform.[9] In China, this led to a prolonged process of sector-making, in which China's villages and cities were made to conform ever more closely to imagined ideal types. The process began in the Nanjing decade (1927–37), under a Nationalist government that fervently believed in the necessity of a planned transformation of the Chinese economy.[10] It reached its apogee in the 1960s and 1970s, when city and countryside had evolved into administratively separate realms, governed by different sets of regulations; all rural people, regardless of occupation, were classified as peasants; and almost all remaining social and economic links between the spheres were severed. Two anecdotal observations from Sichuan illustrate the depth of the gap that emerged in these years: in the late 1980s, rural men at age 18 were on average eight cm shorter than their urban counterparts, due to poorer nutrition—a fact that marked them as people of a different type, immediately recognizable on the rare occasions they came to the cities.[11] And in the 1990s, rural people in Jiajiang burned imitation urban household registration cards as offerings to their deceased relatives in the hope that this would save them from being reborn as peasants.

A quarter-century of market reforms has eroded some of the barriers that separate rural and urban worlds but left others in place. Rural-to-urban migration on a truly massive scale (estimated at 140 million in 2003, or 10 percent of China's population) coexists with the systematic exclusion of migrants from citizenship in the places where they live. The household registration (*hukou*) system, originally conceived to prevent migration, now serves to keep migrants permanently uprooted in their new place of residence, denying them access to healthcare, education, and other services funded by their work.[12] Since the mid-1990s, the central government has repeatedly announced its intention to abolish the *hukou* system, but most municipalities have already issued their own blatantly discriminatory regulations in anticipation of the reform. As late as 2008, experts concluded that the "invisible walls" that divide rural and urban China are still in place.[13] Moreover, the rural-urban separation no longer rests primarily on administrative rules but on the rhetorical construction of rural people as quasi-ethnic aliens, who must be tolerated in the cities for economic reasons but cannot be absorbed into the urban population. Central to this view is the discourse on *suzhi*, or "quality"—defined, in a circular fashion, as the positive personal qualities that China's rural masses lack. The people who help build and sustain China's cities and whose labor underpins the lifestyles of the urban middle classes are thus permanently fixed at the bottom of the hierarchy of value.[14]

The Skillful Peasant

The rural-urban gap is produced and sustained by a model of rural people as frog-in-the-well peasants who live essentially local lives, whose main bonds are with a territorially defined community and with the land they work, who do not participate in regional or national exchange networks, and who are therefore unqualified to participate in public life.[15] This model can be traced to the antitraditional and antipopulist iconoclasts of the May Fourth generation, for whom China's rural population was "a culturally distinct and alien 'other,' passive, helpless, unenlightened, in the grip of ugly and fundamentally useless customs, desperately in need of education and cultural reform."[16] For the May Fourth reformers and their intellectual heirs, peasant lives were essentially reactive, adaptations to varying (but invariably harsh) local conditions. Commerce itself was a hostile force, alien to the mindset of self-sufficient peasants who were fearful and ignorant of the outside world and lost out in every encounter with wily urban merchants.[17] Although never accurate as a description of rural reality, and totally inaccurate in post-reform China, when millions of rural people participate in the industrial and post-industrial economy of China, the model of the earthbound, rooted peasant persists to the present day. Consider the following statement in a popular ethnography, published in 2004, of an unnamed village:

Peasants feel a strong attachment to the soil; the soil is their home; they themselves are like rice plants and the soil is the ground out of which they grow and the home to which they return. Old peasants find it difficult to get used to life in urban apartment blocks. The reason is curious: "We're not used to living on the upper floors; we can't soak up the *qi* [smell or life force] of the soil there"—in other words, they cannot live day after day directly on the soil and therefore want to leave. . . . In fact, peasants are plants; they are soil; they are a cyclical return outside time and history.[18]

In calling peasants plants, the author does not mean to disparage them; quite to the contrary, he sees their "rootedness" as proof that they live more natural lives. In ways that echo nineteenth-century European depictions of the Occident, peasants are here used to represent a way of life less alienated but also more fixed and static than that of city dwellers—a realm of necessity, contrasted with the urban realm of awkward freedom and problematic choice.[19] This contrast, I insist, is an illusion. The link between a farmer and the land is no more direct and "organic" than that between any other skilled producer and the materials with which he or she works. It is necessarily mediated by skill and knowledge, which are essentially human, social categories. Land in itself has value only insofar as it can be made to yield products, a process that requires skill; these products in turn have value only insofar as they can be turned into consumable or marketable goods, and this, too, requires skill. Social life in the countryside (as anywhere else), I argue, is centrally concerned with the *production and reproduction of economically useful skill*, because economic activity—regardless of its location and resource base—is impossible without skill.

Social Organization and Skill-Producing Groups

François Sigaut, a French anthropologist of technology, argues forcefully that all social organization is at least partially about the production of skill:

From a technological point of view, the skill-producing group is a basic social unit present in all societies because a society without techniques is inconceivable. This unit can take a wide variety of forms and enter into extremely diverse combinations with other units such as the family, the residence group, the age-group, etc. All this is a function of such factors as the kinds of skills concerned, the social values placed on them, indigenous ideas on learning, the distribution of activities by rank and gender, and so on. Ideally, the morphology of all societies ought to be reconstructed from scratch, taking into account this necessary but, until now, unnoticed unit. We are obviously far off the mark.[20]

Seeing skill reproduction as a central function of social groups can indeed alter our understanding of social organization. Take kinship, for example: Chinese kinship is often viewed as centrally concerned with the organization of agnates for ritual and political action and with the accumulation of corporate resources, typically land. A moment's reflection will show that the protection of local bodies of knowledge, the transmission of such knowledge to younger generations, and the allocation of tasks by gender and generation are among the more important activities of many Chinese kinship groups. As I show in Chapter 3, most Jiajiang papermakers live in communities composed of agnatically related men and their families. Kinship and technical competence overlap so much that people typically defined the skill-bearing community in kinship terms: "People of our family have been making paper for generations" (*women jiazu zuzu beibei caoguo zhi*). Yet the kinship that one encounters in the paper workshops differs markedly from Chinese kinship as described in the literature.[21] Unconcerned with claims to land and status, practical kinship among papermakers centers on the establishment of work relations and the management of information flows within and among workshops.[22] Rather than stressing vertical descent along patrilines, papermakers emphasize horizontal bonds between men of the same generation and mutual obligations between men of junior and senior generations. Inclusive kinship practices not only result in an easy flow of know-how among agnates but also reinforce boundaries between kin and non-kin, and thus keep knowledge in the hands of the kinship group.

Although the practices I observed in Jiajiang can be shown to serve useful functions in the context of the paper industry, I am not certain that they evolved in response to concrete technical needs, or even that kinship practices among papermakers differ from those of their farming neighbors. It may be that practices that encourage cooperation and knowledge sharing among relatives are widespread but often go unnoticed because they do not fit the view of Chinese kinship as concerned chiefly with juridical claims to land and power and as characterized by competition among different descent lines. Looking through the lens of skill may help us to see patterns of social organization that remain hidden as long as we think of rural people as land-based peasants.

Models of Peasantness

The argument that social integration arises from functional differentiation has been made most clearly by Émile Durkheim, but it can be traced to Aristotle: "The state consists not merely of a plurality of persons, but of persons who differ in type; for a state does not come from people who are alike."[23] The division of labor, interdependence, and exchange create community; "fixity and same-ness of conditions"—Marx's description of peasant life—isolates people from one another and renders them unfit for public life. Specialization and exchange were widespread in late imperial China, where most industrial goods were produced in the countryside by specialized or semi-specialized artisans. In recognition of this fact, late imperial and early Republican governments were generally supportive of rural craft industries. Specialization was seen as a necessary element in a mixed rural economy; crafts and sidelines were seen as contributing to a stable social order, because they generated income that allowed land-poor farmers to stay on the land. Specialization was to be discouraged only when it interfered with the agrarian economy by drawing labor off the farm or when it led to dangerous concentrations of unruly male wage workers[24]—and even this could be tolerated if the fiscal or commercial payoff was large enough. Madeleine Zelin estimated that the great saltyards of Zigong, about one hundred kilometers from Jiajiang, employed between 68,000 and 98,000 workers, which made them one of the largest concentrations of industrial workers in the nineteenth-century world.[25]

Specialization did not create "citizenship"—the notion did not gain currency in China until the 1890s, and even ardent republicans thought of rural people as citizens only in the most abstract sense. However, specialization linked rural people to an economy of goods and signs that stretched from remote villages to the centers of power.[26] Papermakers in Jiajiang were always aware that they produced not just a useful good but a symbol of China's literary and bureaucratic culture. For over three hundred years, their tribute paper (*gongzhi*) was used in the provincial civil service examinations, and even after the abolition of the examination system in 1905, national and provincial governments took an active interest in the industry because they needed paper. Paper, of course, lends itself to the construction of cultural ties better than many other goods, but all goods are to some extent imbued with meaning and can be used as vehicles for social or cultural claims. Weaving, for example—by far the most important rural craft—was associated with gendered divisions of labor, the moral order, and social stability, and craft producers could invoke such norms in defense of their craft. As we shall see in Chapter 4, papermakers in Jiajiang used cultural claims to attract the attention of provincial and sometimes national elites and to lobby for tax breaks.

The vision of local specialization as normal, necessary, and positive began to change in the first decades of the twentieth century, due largely to the conviction among Western-educated elites that China needed to industrialize rapidly in order to defend itself against the West and Japan. In the early twentieth century, industry came to be seen as a means to "save the nation" (*yi shiye jiu guo*) in its Darwinian struggle for survival. At the same time, urban elites came to think of rural people as peasants who were too short-sighted or ignorant to be left in charge of crucial national resources. As Henrietta Harrison shows in her study of Shanxi in the closing years of the Qing and under the Republic, Western-educated elites were so intent on promoting "Industry" (in the sense of mechanized factory production) that they overlooked the vibrant industries that actually existed, and at times even actively suppressed them in favor of (often less successful) modern-style operations. Government hostility to small-scale industry led to the concentration of manufacturing in the cities and the agrarianization of a previously mixed rural economy.[27]

As much recent research shows, there was a basic continuity between pre- and post-1949 development strategies.[28] Like their Republican predecessors, the Chinese Communist Party (CCP) pursued a strategy of "commanding heights" industrialization, in which all efforts were concentrated on large-scale, modern, and predominantly urban industries, in particular defense industries. Operating in a hostile international environment and under conditions of great scarcity, CCP economic planners created a bifurcated economy in which the rural sector was subordinate to an encapsulated and protected urban sector. The features of this system are well known: from 1953 on, a system of compulsory purchases at state-fixed prices pumped cheap grain, cotton, and other inputs into the urban sector, while prices for farm inputs and consumer goods were kept high, ensuring steady profits for state factories. In order to prevent rural people from migrating to the cities and thus diluting the gains of modernization, the state tied rural populations to their natal or (in the case of women) marital villages. Migration and diversification, common routes to economic success until the 1950s, were curtailed and eventually banned. In contrast to urban residents, who had access to state-guaranteed subsistence and sometimes substantial welfare packets through the *danwei* (work unit) system, the livelihood of rural people depended on local resource endowments and the vagaries of the weather. Much has been made of the Maoist policy of "walking on two legs," that is, the parallel pursuit of agriculture and small-scale industry. Yet the objective of this policy was to *enhance*, not to reduce, rural self-sufficiency. The Maoist ideal for the countryside was the self-reliant, insular collective that produced surplus grain and other inputs for cities but required nothing from the urban sector. The commune and brigade enterprises of the 1970s, precursors of the post-Mao rural industrial boom, were designed to "serve agriculture" and were expressly forbidden to compete with state-owned enterprises for raw materials, capital, or markets.

My point here is not simply that rural people were materially disadvantaged relative to urban people—although of course they were, with urban incomes and consumption levels in the Maoist years two to three times higher than rural levels[29]—but that they were differently integrated into the body politic. Most urban people belonged to work units that were highly specialized and firmly incorporated into territorial and functional hierarchies. Since the state planning system kept duplication to a minimum, most work units were unique within a given territory: each province or prefecture had one, and only one, ball-bearing plant or construction company. Work units were also integrated along functional "lines" (*tiaotiao*, as opposed to territorial "blocks," *kuaikuai*), in such a way that each unit depended on upstream and downstream units in the same administrative system. In this complex and rigid structure, problems in one place could easily reverberate throughout the entire system. Each stoppage or slowdown because of problems somewhere in the supply chain was an object lesson in interdependency that must have brought home to workers a sense of their own indispensability.[30] By contrast, by suppressing specialization and turning villages into self-sufficient cells, Maoism removed rural people from the web of mutual dependency and exchange of which they had been part. Peasants as a class were still needed and in an abstract sense "revolutionary," but each individual peasant stood in the same unspecific relationship to the whole: unconnected to people outside the village community, he or she faced only in two directions, down to the soil and up to the state. If peasants differed from one another, it was only because they adapted to different local conditions, in the same way that a cabbage grown on sandy soil differs from one grown on loam. This view of rural people as self-sufficient peasants continues to shape the perception of them among urban people and is used to justify their exclusion from full citizenship rights.

Two Types of Deskilling

Like other resources—land, water, factories—skill is contested and subject to distribution struggles. Although it cannot be expropriated in quite the same way as tangible assets, it can be monopolized—or, to the contrary, lost, stolen, or destroyed. A tradition stretching back to Charles Babbage and Karl Marx links the advance of capitalism to the breaking up of complex production processes into shorter, simpler ones that can be carried out by unskilled (typically female or child) labor or machines. In a craft workshop, most work is performed by skilled labor. Capitalist factories, by contrast, save costs by dividing the production process and purchasing exactly the quantity of skill needed for a given task: skilled labor for machine set-up, unskilled women for milling, child labor for cleaning, and so on. This, rather than the increased efficiency and speed that results from repeating the same action all day, is the motive for the minute subdivision of tasks in capitalist industry.[31] Marx's comments on deskilling were amended by Harry Braverman, who focused on the progressive separation of creative tasks (design, planning, and the like) from execution. The result, Braverman argued, is the inexorable "degradation of work" under capitalism.[32]

Braverman's work spawned a large number of studies, most of which found a much more complex relation linking capitalism, technical change, and the labor process. On one hand, critics argued that deskilling cannot have been as dramatic as Braverman contended, because *there is little skill to begin with*. In this view, skill in the sense of individual competence or control does not exist or is irrelevant to both industrial and preindustrial work.[33] As feminist historians have long pointed out, skill is at least in part a social construct: a claim made by powerful groups of workers (often male, unionized, and white) to exclude competitors (typically women and migrant workers).[34] Which work counts as skilled often has more to do with *who* performs it than with the complexity of the work itself, and what appears as deskilling may in many cases have been a simple leveling of the playing field. In direct opposition to this view, some critics argued that deskilling is less grave than Braverman thought, because *skill is constantly reproduced in the labor process.* Workers, it can be shown, resist deskilling, and their resistance shapes managerial decisions.[35] More generally, technological progress creates new skills at the same time as it destroys old ones.[36] Sigaut even argued for a general "law of the irreducibility of skills," because the "constantly renewed attempt to build skills into machines . . . is constantly foiled because other skills . . . develop around the new machines."[37]

"Bravermanian" deskilling in the capitalist factory can be contrasted with what could be called "Scottian" deskilling, after James Scott, its most astute analyst. Scott and other students of the postcolonial global South have focused on deskilling among subaltern groups—peasant farmers, small artisans, and indigenous peoples— rather than on the factory proletariat. The agents of deskilling in this case are not capitalists but intellectual visionaries, state technocrats, colonial administrators, and other agents of modernizing states. The dispossession of subaltern groups is the result not of the capitalist profit motive, although this may also play a role, but of the pursuit of a modern vision of a world liberated from material want. This process began in Enlightenment Europe, when new information-processing technologies made it possible to extract "blind" knowledge from "the ateliers and the hands of the artists," where it lay buried, and to circulate it in print. Cynthia Koepp has shown how the careful reconstruction of artisanal knowledge in Diderot and d'Alembert's *Encyclopédie* hid a "subtle and comprehensive expropriation of nonliterate knowledge by the literate culture, an attempt, largely successful, to remove the inefficient and inarticulate world of work from the hands and mouths of the workers and to place it in printed form before the eyes of an enlightened 'management' whose ordered purposes it would serve."[38]

The French *encyclopédistes* mark the beginning of a growing concern with the world of work as the key to greater prosperity and thus greater happiness for all. Productivism—the notion that a better world can be achieved by unleashing mankind's almost infinite potential for increased production of material goods—informed liberal, socialist, and fascist policies in Europe and America. Nineteenth- and early twentieth-century industrial reformers saw social ills as the result of ignorance, inefficiency, and waste, rather than of injustice, and aimed to restore social harmony through the systematic reorganization of work along rational, scientific lines. On both sides of the Atlantic, scientists and managers looked at the human body as a machine capable of vastly improved output, if only properly supervised and instructed (the concern of F. W. Taylor's "scientific management") or properly fed and rested (the concern of the European "science of work").[39] The same utopian belief in the necessity and possibility of a planned reorganization of work can be found in the "Soviet Taylorism" of Aleksei Gastev's Central Labor Institute,[40] in the radical antitraditionalism of postcolonial regimes in Africa and Asia,[41] and in the impact of Taylorism and Fordism on Chinese management, both before and after 1949.[42]

Jiajiang papermakers fought over skill at many different levels. In household workshops, men excluded women—including their own wives and daughters—from certain operations to prevent them from acquiring crucial skills. In the large proto-capitalist workshops that sprang up during times of expansion, employers kept a tight grip on production technology. My focus in this book is on a process of state-led skill expropriation that began tentatively in the 1920s and culminated in the campaigns and struggles of the 1950s and 1960s. "Socialist deskilling"—the massive onslaught on traditional handicrafts during the Maoist years—bears the imprint of both forms of skill expropriation that I outline here. A rationalizing, profit-maximizing impulse—to get "more for less"[43]—is present in attempts by socialist managers to reorganize the paper industry. Yet, more important, we are dealing here with state actors and elites that sought to impose a radically new regime of knowledge, one in which technological control lay in the hands not of "unenlightened," "fractious," "selfish" local people but of experts who spoke and acted in the name of the nation. Even though skill extraction and redistribution were pursued in the name of science and rationality, they were not necessarily rational. As William Kirby has argued in a different context, such policies were motivated by a belief in the profound irrationality of all historically grown structures ("remnants of the feudal past") and an almost religious belief in "scientific" planning.[44]

The Nature of Skill

Skill matters because most societies distribute income, wealth, and power on the basis of real or assumed competence, in other words, of skill. Yet there is surprisingly little agreement on what skill is and where it is located. Skill has been seen as a "thing," a property or possession of the skilled artisan or worker, but it has also been described as little more than a discursive claim. Skill has, with good reason, been characterized as a form of "personal knowledge," securely and inalienably inscribed in individual bodies. At the same time, it has been depicted as a property of social groups rather than individual actors. These views of skill as "real" or discursively constructed, physically *embodied* or socially *embedded*, are not mutually exclusive, yet some conceptual clarity is needed if we want to understand what happens when individuals or communities are deskilled.

My understanding of skill is shaped by two bodies of literature— phenomenological philosophy and cognitive science research—that appear quite remote from the concerns of social historians. Phenomenologists such as Martin Heidegger and Maurice Merleau-Ponty have long argued that skill is central to the human condition. Our primary mode of being in the world is not ratiocination but skilled, active, bodily engagement with our immediate surroundings. We do not usually contemplate external objects as detached subjects but respond spontaneously to the opportunities for action that our environment affords. It is only when the flow of activity breaks down (in Heidegger's well-known example, because the hammer that we are using to drive a nail into the wall breaks) that we become aware of the "objective" qualities of our environment—say, of the hammer as a hard object of a certain size and shape—and of our own existence as detached subjects.[45]

What Heidegger intuited but did not prove has been fleshed out by scientists working at the interface of neuroscience, robotics, artificial intelligence, and philosophy.[46] Intelligent action does not require central processing taking place in a bounded mind but emerges from distributed "cognitive and computational processes that are busily criss-crossing the boundaries of skin and skull."[47] By building ever more extended feedback loops that extend from our brains through our bodies to body extensions (tools) and the external world, we literally think with and through our bodies and the environment. Most mental processes involve "scaffolding"—the use of external features in the physical or social world that augment and constrain the problem-solving capabilities of the biological brain and enable it to solve problems that cannot easily be solved by the unaugmented brain.[48] We "scaffold," for example, when we arrange tools on a workbench in such a way that they serve as prompts for action, or when we use pen and paper to multiply large numbers.[49] More generally, language, culture, and institutions can be seen as scaffolding devices that reduce computational demands on the brain by "offloading" information onto the external world, from which it is accessed in ready-made form whenever it is needed.

The feedback loops we build can include other people; in fact, much cognitive work is socially distributed, resulting from the structured interactions of people with one another and with artifacts in the world.[50] As the anthropologist and cognitive scientist Edwin Hutchins has argued, we give too much epistemological credit to the individual mind. Cognition "in the wild" (Hutchins's example is the determination of the position of a naval vessel) is often distributed across networks of actors, and intelligent performance emerges not simply from minds or groups of minds but also from the culturally constructed ways these minds are linked into flexible and robust systems: in his example, from the spatial positioning of officers on the bridge, the division of tasks among those officers, the use of standardized signals, and so forth. In a similar vein, Jean Lave and

Etienne Wenger have argued that the proper sites for understanding the production of practical, everyday knowledge are "communities of practice"—small, informal groups held together by the sustained pursuit of a shared enterprise.[51]

Skill, then, is located not in the encapsulated mind, not even in the body as a repository of "embodied knowledge," but at the interface between the skilled person and his or her surroundings. It is "a property not of the individual human body as a biophysical entity . . . but of the total field of relations constituted by the presence of the organism-person, indissolubly body and mind, in a richly structured environment."[52] This view of skill as distributed across a field of relations, rather than safely stored in individual persons, has implications that become clear later in this study. It explains, for example, how Jiajiang papermakers could recreate their craft in the 1980s, after a hiatus of almost twenty years, from information dispersed throughout their natural, social, and symbolic environment: among other things, from the bodily memory of aging practitioners; from the layout of bamboo forests, sheds, and tools; from shared assumptions about who does what kind of work, how to organize work teams, how to allocate tasks by gender and generation, and how to cooperate with kin and neighbors. All this was underpinned by a shared, unarticulated, matter-of-fact understanding of the social and symbolic world, which guided papermakers in the re-creation of their trade. This wide dispersal of information across heterogeneous media also explains why the state never succeeded in transplanting Jiajiang papermaking skills to places outside Jiajiang: despite active support from papermakers, their skills could not be reproduced outside the social, cultural, and material support structure that existed in Jiajiang.

Skill as I understand it shares many features with Pierre Bourdieu's concept of *habitus*, defined as an ensemble of acquired, durably installed bodily dispositions that guides our everyday practices and enables us to act in accordance with what is objectively possible under the specific social and historical conditions that generate a specific *habitus*. *Habitus*—described by Bourdieu in terms of expertise, virtuosity, improvisation and, indeed, skill—is the "generative principle" that makes it possible for us to think and act in ways that are "as remote from creation of unpredictable novelty as . . . from simple mechanical reproduction." Like skill, *habitus* is simultaneously knowledge and practice, a form of spontaneous mastery in which the virtuoso practitioner "carries and is being carried" by his or her performance, "like a train laying its own rails."[53] Both *habitus* and skill are nondiscursive and are acquired through mimetic learning; both are embodied principles of "regulated improvisation" rather than collections of fixed rules.[54] Like skill, *habitus* is not the possession of any one individual: although individually embodied, it is generated through collective practice and instantiated only in social practice, when a similarly engendered *habitus* causes people to act in mutually comprehensible and compatible ways.

For historians, of course, skill is a useful concept only to the extent that it helps elucidate historical change. To what extent is skill a historical category, and what do we gain by looking at historical processes through the lens of skill? The answer is, I believe, that skill constitutes a sort of meeting ground between "big" historical processes—war, revolution, industrialization—and the concrete experience of everyday life. Routine skillful coping with the world around us forms both the silent backdrop from which all conscious thought and action spring and the ground of all individual or collective agency. Jiajiang papermakers did engage with the revolution at a conceptual level, for example, by learning to speak the language of land reform and class struggle (despite the fact that such terms as "landlord" and "poor peasant" were ill-adapted to their situation), but their engagement was primarily as involved actors drawing on a repertoire of previously acquired skills to react to concrete changes in their daily life. Like *habitus*, skill can be seen as "embodied history, internalized as second nature"—a structure of dispositions that enables one to act with a great degree of freedom in certain ways while precluding actions outside one's acquired social and practical competence.[55] *Habitus* is often thought of as a constraint—a fixed horizon of expectations that predetermines our thoughts and actions and leads to ill-adjustment (*hysteresis*) when we rise beyond our assigned class positions or are uprooted by rapid change.[56] In Jiajiang, too, skill, as a system of embodied orientations, can limit opportunities and fix people "in their place"; this is particularly the case for

women, whose embodiment of skill is complicated by harsh demonstrations of male power that make it difficult for them to develop
a sense of competence and control. Yet more often, rapid, state-
induced change throws up material and cultural fragments—
production techniques, organizational formats, ideological justifica-
tions—and people, acting from a background of previously acquired
skills, incorporate these fragments into existing repertoires in ways
that extend their material and social reach. For example, former
wage workers learned to speak in public, organize meetings, and direct work teams, skills that dramatically lengthened the feedback
loops linking them to their material environment and other people.
At the same time, previously privileged groups experienced a dramatic shortening of their social and physical reach and a concomitant loss of competence, autonomy, and power.

Fieldwork and Sources

The data for this book were collected between 1995 and
2004, with the bulk of the field research taking place in 1995–96
and 1998, with shorter return visits in 2001 and 2004. Most interviews were conducted in the village of Shiyan, where I lived in the
guesthouse of a village-owned factory for three months and visited
on a daily basis for another four months. I also conducted interviews
in Bishan village, adjacent to Shiyan, and Tangbian village in the
remote mountain township of Huatou. Most interviews took place
in the presence of a local guide (the chairwoman of the village
women's federation, the village head, and two retired male village-
level cadres taking turns). The social distance between village cadres
and other villagers is so small that I do not think that the presence
of my guides significantly shaped the responses to my questions. In
any case, villagers—in particular former cadres—were quite outspo-
ken in their criticisms of past and present government policies.[57]
Most of the time, I was also accompanied by a researcher from the
Institute of Rural Economics of the Sichuan Academy of Social Sciences, who translated dialect expressions into standard Chinese until I had a working knowledge of rural Sichuanese.
    Fieldwork in Jiajiang, though enjoyable, was not always easy: the
county had only recently been opened to foreign visitors and several

factors (among them the presence of an army garrison and a secret nuclear research institute in the county) made the local authorities less than enthusiastic about my presence. Moreover, manual paper production technology had been declared a "state secret at the district level," which made it technically illegal to discuss papermaking technology with foreigners. I heard several stories about Japanese or Taiwanese "spies" who had done undercover research in Jiajiang. In the one case I was able to ascertain details, the spies were in fact scholars from Chengdu, doing research for the Taiwanese folklore studies magazine *Hansheng*.[58] Fortunately, the papermakers did not share this concern for secrecy, perhaps because they understood more clearly than did local officials that one does not learn a craft by interviewing its practitioners. Interviews took place in workshops and courtyards and were open-ended, sometimes meandering, with friends and neighbors joining long discussions that ranged from production processes to local gossip. One of the great advantages of discussing the concrete details of daily work, I found, is that it allowed me to treat my informants as skilled actors competent in all areas of their daily lives. Much social science research defines its field of inquiry in ways that makes the outside expert appear more knowledgeable than the local informant. Shifting the emphasis to a field in which informants were highly skilled allowed me to partially redress that imbalance.[59]

Oral history interviews were supplemented by written sources from a variety of archives. Most of the holdings of the Jiajiang County Archive were destroyed during the liberation of the county in 1949 or during the Cultural Revolution. In Chengdu, however, I had access to the archives of the Provincial Reconstruction Bureau (covering the years 1936 to 1949) and of the Ministry of Industry of the Southwest Administrative Region (covering 1949 to 1952). I was also given permission to use the rich collections of Republican-period journals and newspapers of the Sichuan Provincial Library and Sichuan University. Together, these sources give a detailed picture of Jiajiang papermaking from the mid-1930s to the 1960s. No material of comparable quality is available for earlier periods, but stone inscriptions, hand-copied from tombs and temples in the Jiajiang hills, throw light on some aspects of social organization at that time.

Organization of the Book

Chapter 1 opens with a description of manual papermaking technology, the gender and generational division of labor, and the developmental cycle of household workshops (which, like farms, expanded and contracted along with the demographic composition of households). Next, I discuss the weft and warp of skill reproduction, that is, the vertical transmission from one generation to another and the horizontal sharing of skills among relatives and neighbors. In Chapter 2, I discuss the overlapping markets for raw materials, grain, credit, labor, and paper that underlay pre-1949 paper production, as well as the guilds, temple associations, secret societies, and other intermediate organizations that linked local papermakers to the market and the state. Chapter 3 deals with the place of Jiajiang papermakers in the political economy of late Qing China and the growing pressure to modernize craft industries in the Republic. In Chapters 4 and 5, the focus shifts to Shiyan village during the land revolution and collectivization. From 1950 on, papermakers were classified as peasants and participated in all the rural movements and campaigns. At the same time, large numbers of papermakers were supplied with state grain in exchange for paper. Collectivization saw a consolidation of paper workshops into larger units and a sharp decline in the number of people entitled to grain rations. It also saw a gradual deskilling of papermakers, as officials in the Second Light Industry administration extended their control over the paper workshops.

Chapter 6 describes the Great Leap Forward and the subsequent famine, which wiped out one-quarter of the population of the paper districts and all but destroyed the industry. In order to survive, papermakers began to cut down their bamboo and planted maize and sweet potatoes on the steep, quickly eroding slopes. After the famine, teams that continued to produce paper were accused of "eating guilty-conscience grain" (*chi kuixinliang*) and ordered to become grain self-sufficient farmers. Nonetheless, papermaking continued, as growing numbers of individuals and teams began to sell paper to black market traders who have plied the markets of Sichuan since the mid-1970s. The early development of the black market in Sichuan gave Jiajiang a head start in the post-Mao years, which are the subject of Chapters 7 and 8. With the legalization of private trade and the burgeoning demand for handmade paper, Jiajiang traders began to travel throughout China, retailing directly to artists' associations, schools, and department stores. By the 1990s, there was hardly a major city in China without a paper outlet from Jiajiang. The revival of the industry was accompanied by a minor technological revolution, as papermakers adopted machines and aggressive chemicals that drastically reduced production time, and by the rise of a two-tiered production structure, in which extremely small household workshops struggled, often in vain, to join the ranks of established, labor-hiring producers of calligraphy paper. Finally, Chapter 9 focuses on attempts by the people of Jiajiang to reconnect the social tissue that had become thin and fractured during the Mao years and to repair the structures that underpinned the reproduction of skill.

1   Locations of Skill

TANGBIAN VILLAGE CAN BE reached only by a footpath that leads from the sleepy market town of Huatou to a perpendicular sandstone cliff. A narrow cleft in the cliff is the only access to the plateau on which Tangbian lies, and although steps and handrails are cut into the stone, the passage is difficult, especially on rainy days. Once the cliff is traversed, the scenery becomes idyllic, with tea gardens, bamboo groves, and whitewashed houses scattered along creeks and ponds. Xia Deli and his family of seven live in a wooden house, with living quarters and papermaking sheds arranged around paved courtyards. The Xias pay a high price for their idyllic location: every year, about fifteen tons of raw materials have to be carried up from Huatou, and two to three tons of paper carried down. Porters who carry more than their body weight in coal or soda on their backs earn 15 yuan per trip, or 30 yuan a day—three times the going rate for unskilled labor at the time of my visit in 1996.

In 1996, the workshop employed eight people: Xia Deli and his wife, their daughter, her husband, and four hired workers. Deli's main tasks were management—he bought raw materials, sold the finished paper, and supervised his workers—and pulp preparation. His wife, Ms Shen, worked the family's 0.5 ha of land, looked after two grandchildren, raised five pigs and several sheets[1] of silkworms every year, and cooked three meals for ten people every day. The couple's daughter and son-in-law helped with pulp preparation and did most of the finishing and packing. As in other big workshops, the core processes of papermaking—molding paper sheets and brushing them onto drying walls—were left to hired workers. The Xias

23

employed two couples in their fifties, both from Hedong (East of the River), the more developed and accessible hilly part of Jiajiang to the east of the Qingyi River. They were paid piece-rate wages (7.6 *yuan* for one thousand sheets for the male vatmen, 4.5 *yuan* for the female brushers) and given board and lodging. Local custom dictated that workers were given three daily meals of rice and vegetables, with meat added every second or third day. In addition, men were given a pack of cheap cigarettes every day and a bottle of 120-proof liquor every four or five days. Workers ate with their employers, although they tended to eat quickly and return to work after gulping down their food.

In contrast to papermakers in Hedong, who have long since abandoned their traditional wooden steamers, papermakers in mountainous Hexi (West of the River) still use the old steaming technology. These steamers consist of three parts: a ring wall, about two meters high and three to four meters in diameter, with an opening for fuel; a huge iron pan that rests atop the wall; and a conical wooden "pot," about three meters high and three meters wide at its base, similar to the rice steamers traditionally used in Chinese kitchens (Fig. 2). The large steamers used in the collective period held fifteen tons of fresh bamboo; the Xias' steamer, smaller but still impressive, holds seven tons. On average, the Xias steam three full "pots" every two years: a full, seven-ton load after the summer bamboo harvest, and a reduced load in winter or spring. Unused steamers quickly rot in the humid climate of Sichuan, which is why the Xias allow neighbors to rent their steamer for a small fee when they do not need it themselves.

By the 1990s, workshops like Xia Deli's had become rare. Papermakers in Hedong had replaced wooden steamers with steel-and-concrete pressure cookers that reduced labor input and turnover time (see Chapter 7 for a discussion of these changes). In the Hexi mountains, however, the production process had changed little since the early twentieth century. Even here, however, technology was not stagnant: since the 1980s, coal had replaced bamboo as steamer fuel, caustic soda had been substituted for potash, and pulping had been mechanized.

Papermaking Technology

Papermaking is a complex and demanding craft with an elaborate division of labor. Papermakers speak of 72 separate operations (nine times eight: a lucky number), and detailed descriptions of the industry list 20 operations, each consisting of several smaller steps.[2] For our purposes, papermaking can be divided into two main parts: "steamer work," which includes the seasonal processes that transform bamboo or other fibrous matter into *liaozi* (partially broken fibers), and "vatwork," the year-round operations that transform *liaozi* into pulp and pulp into paper. "Steamer work" begins with the cutting of fresh shoots in May or June, when the bamboo fibers are long and elastic; if harvested later, the fibers become short and brittle. Bamboo responds well to fertilizer and labor input, and bamboo stands near houses are often surrounded by low banks and kept free of weeds. Most bamboo, however, grows in the hills, too far from homes to be fertilized or weeded. The two bamboo species most commonly used in Jiajiang, *baijiazhu* (*Phyllostachys puberula makino*) and *shuizhu* (*Bambusa breviflora munro*), have vegetation cycles of fifty to sixty years; at the end of each cycle, the bamboo flowers and the central rhizome (called the *magen*, "horse root") dies. Surviving side roots form new rhizomes, and after a few years, yields return to normal. This process is called "changing the head" (*huan tou*). The years in which bamboo flowers can devastate paper workshops, but, contrary to a common belief, all bamboo plants of the same species do not necessarily flower at the same time, and not all plants that flower die. Paper workshops that lose their own supply of fresh bamboo can switch to imported *zhuma*—dried and split bamboo—which is readily available at local markets.

Cutting the bamboo requires considerable experience, since fixed portions of first-, second-, and third-year bamboo have to be cut each year to ensure continuous regeneration. After the harvest, the fresh bamboo is split, cut to length, and soaked in water (Fig. 1). In the summer heat, the outer skin of the bamboo rots away until only the fibrous "meat" (*rou*) of the stalks remains. After several weeks, the bamboo is rinsed, mixed with quicklime, and soaked again for two weeks. It is then packed into steamers (Fig. 3) and boiled for six or seven days. After letting the steaming mass cool for a day, a group of men climb atop the steamer and begin to pound the *liaozi* with long pestles. The fibers, loosened by the heat and quicklime, must be separated while still hot; otherwise, the dissolved lignin hardens and the fibers stick together. Layer after layer, the *liaozi* is raked out of the steamer and spread out on a paved area, where it is beaten with long clubs or hammers (Fig. 4). Next, the fibers are rinsed in ponds or mountain streams (steamers are always built close to water) to remove the lime and lignin. The *liaozi* is then packed back into the steamer and cooked a second time, this time with potash or soda ash to form lye water. After another five days of steaming, it is once again rinsed in water. As the lye is washed out, the color leaves the fibers, and the *liaozi* turns white and fluffy like cotton wool. It is then piled into densely packed "cakes" and left to ferment for a few weeks.

In contrast to the seasonal work of steaming, vatwork—pulping, molding, brushing, and finishing—continues almost without interruption throughout the entire year. Every morning, the pulper cuts a lump of *liaozi* from the cake, mixes it with water, and treads it with bare feet for about an hour. After the first washing, chlorine bleach is added, left to work on the fibers for about an hour, and then rinsed out in repeated washings. At this point, the fibers are still too long and thick for papermaking and need to be pulped. This was traditionally done in foot-operated tilt-hammer mills (*duiwo*), similar to the mills used to polish rice. By stepping on the lever of a wooden hammer, the pulper lifts the hammer head; when he releases the lever, the hammer falls into a stone trough filled with *liaozi*. This is repeated again and again until the *liaozi* is reduced to pulp (Fig. 6).

The pulp is then brought to the molding shed, an open structure with a large, rectangular sandstone vat in the middle. The vatman adds one or two ladlefuls of pulp to the water in the vat and adds a plant extract to prevent the fibers from clotting. The mixture is then vigorously stirred until it has the color and texture of milk. Paper sheets are formed with a paper mold consisting of two parts: a flexible screen made of thin strips of bamboo, coated with lacquer and tied together with silk thread, horsehair, or fishing line, and a rigid wooden frame to support the screen. The vatman grasps the mold with arms outstretched, bends over the vat, and dips it into the water (Fig. 8). He trails it through the water and lifts it out horizontally, with a slight tilt to the right. This allows the water to run off, leaving a thin sheet of fibers on the screen. Next, he scoops up a bit of water with the right corner of the mold and tilts it to the left, to create a second layer of fibers that intermeshes with the first. He then places the mold on the edge of the vat, removes the strip of wood that wedges the screen to the frame, and carries the screen to a table. The paper is then "couched": the screen with the newly formed sheet is flipped over and pressed down; then the screen is rolled off, leaving a soft, wet sheet of paper on the table (Fig. 9). Sheet after sheet is formed, and each sheet is pressed on top of the preceding one.[3] Over the course of a workday, the vatman molds several hundred sheets. The pile of dripping paper is then put into a wooden press consisting of two heavy boards, a lever, and a rope and pulley. Pressure is increased gradually during the evening; if the paper is pressed too quickly, the water trapped in the pile forms bubbles that cause the sheets to burst.

Preparations for brushing begin late at night, after the vatman has taken the paper out of the press. Women—the men have gone to bed by this time—separate the soft sheets with pincers and spread them out on a table, with ten sheets forming a "fold" (*die*). The next morning, these "folds" are carried to special "paper walls" (*zhibi*) on the outsides of houses or to drying sheds built specially for this purpose.[4] Sheets are separated from the "fold" and pasted onto the wall with a stiff brush; in order to save space, sheets are pasted on top of each other, with ten sheets forming a *diao* (Fig. 10). After a few days, the paper is dry and the *diao*, which now resembles cardboard, can be taken off. The sheets are once again separated, smoothed, sorted, counted, cut to size, folded, and packed. These tasks, known as "finishing" (*zheng zhi*), are usually done by old men or women in between other tasks (Fig. 11).

Labor Demand in Papermaking

Table 1, based on a report on papermaking technology in the early 1950s, shows that at the time 1,107 workdays were required to make 40,000 sheets (presumably large-format writing paper

Table 1
Labor Input
(One run of a steamer holding five tons of fresh bamboo,
producing 40,000 sheets of paper)

Production steps	Workdays
1. Bamboo harvest (incl. transport)	80
2. Bamboo cutting and crushing	88
3. Filling basin, etc.	17
4. First steaming (incl. repair work and filling the steamer)	44
5. Beating and washing liaozi	179
6. Second steaming (incl. repair work and filling the steamer	30
7. Preparing for fermentation	39
8. Odd jobs	70
SUBTOTAL	547
9. Pulping, molding, and brushing for 40,000 sheets	560
TOTAL	1,107

SOURCE: Sichuan Archives, Gongyeting 1951 [171: 1], 136.

although this is not explicitly stated) from one run of a five-ton steamer. Another report from the 1950s assumes a much lower labor input of 115 workdays per 10,000 sheets, probably because it includes only routine vat work (step 9) but not the work around the steamer (steps 1 through 8). A realistic estimate of the labor expenditure of a workshop should exclude steps 1 to 3, which are typically performed not by permanent workshop staff but by unskilled hired workers. It should include steps 4 to 8, which are performed with workshop labor or with labor exchanged from other workshops. This results in a labor expenditure of 922 days for a steamer of this size, or 230 days for 10,000 sheets. Assuming that a workshop steams 1.5 "pots" a year, total yearly labor demand is 1,383 workdays, equivalent to the fulltime labor input of four to five workers.[5]

Labor demand in manual papermaking is relatively rigid, due mainly to the nature of vatwork. The "vat," as the basic unit of production in a paper workshop, includes not only the molding vat itself but also drying presses, paper walls, and other equipment organized around the vat. At the very least, such a unit needs a pulper, a vatman, and a brusher. However, a vat with only three workers operates below its full potential, because the workers cannot devote themselves entirely to their tasks. Only if they are aided by two helpers who carry loads and help finish the paper can the vat be operated at full capacity throughout the year. This number—five workers—does not include the female labor needed to prepare three meals a day, to wash and mend the clothes of household members and of hired workers, to raise children, and to take care of vegetable plots, pigs, chickens, and sometimes silkworms. Nor does it include the labor needed for repairs, marketing, and transportation. If all this is included, a one-vat workshop needs six or seven able-bodied adult workers, not counting several hundred days of hired or exchanged work during the steaming season. A workshop with seven or more workers could add a second vat, as Xia Deli did in his workshop. There are, however, few economies of scale in manual papermaking, and a two-vat workshop needs roughly twice as much labor as a one-vat workshop in order to make full use of the equipment.

Rigid labor demand around the vat leaves household workshops with two options. Today, as in the past, most workshops are too small to utilize their vats fully and too poor to hire long-term workers. Although almost all workshops hire day laborers for unskilled work and exchange labor with their neighbors, they remain chronically short of labor. Labor-short workshops are forced to "skimp work and reduce materials" (*tougong jianliao*), that is, to cut corners and try to disguise the resulting flaws. Small household size goes hand in hand with intermittent work, poor paper quality, and market strategies that aim at short-term gain rather than long-term trust in the marketplace, a pattern contemptuously described by owners of large workshops as "playing around with paper" (*shuashua zhi*). Continuous, year-round production of quality paper is far more profitable but requires the use of wage labor, not only because the number of workers in a fully staffed workshop exceeds the size of most families but also because the core tasks of molding and brushing require a strict discipline difficult to enforce on household members. Vatmen, in particular, need to work with machine-like regularity, since they set the pace for the entire workshop. They are shielded from all other tasks and given the best food—eggs, meat, fat, and sugar—so that they can work for ten to twelve hours every day. Sons sometimes work as vatmen but are often more interested in learning the tasks they will need as independent workshop owners, such as pulp preparation and management. Sons are also more difficult to discipline than hired workers, who can be dismissed if they do not work hard enough.

Gender and Generational Divisions of Labor

Vatwork is divided by gender and generation; other, more peripheral tasks are less clearly defined in these terms. Of the main tasks at the vat, pulping is the most crucial, because it determines whether the paper will be matte or glossy, smooth or porous, soft or stiff, water-resistant or absorbent, white or grey. Every workshop has its own "recipes," which are kept secret by the household head. Pulping is usually combined with management functions, such as supervision of workers, purchase of raw materials, and selling the paper. Molding requires less experience than pulping but far more stamina and strength. It is strictly male work; even during the Cultural Revolution, when women were told that "they can do whatever men can do," only two or three audacious women in Shiyan learned to mold paper. Brushing the soft, dripping paper sheets onto drying walls is women's work; men brush paper only if they are too old or weak to mold. Wage brushing is less common than wage molding, but women do hire out (and did so in the past), not only to their own relatives and neighbors but also to distant places.

Even by the harsh standards of rural China, work in the paper industry is (and always was) grueling. Vatmen spend ten to twelve hours a day in damp molding sheds, with their hands and arms half-submerged in the vat. Molding continues in the winter when temperatures fall below the freezing point, and stops only when ice forms in the vat. In summer, workshops are filled with mosquitoes and the smell of rotting pulp, and the heat causes the pulp to curdle, making it difficult to mold. Vatmen often spit blood in the winter months, due to the damp and chill in the workshops, to the strain on their lungs when they lift the mold, and to incessant smoking. The mold itself is not heavy, but every time a vatman lifts it out of the water—five hundred to a thousand times a day—he also lifts a water column that weighs about eight kg.

Papermaking in Jiajiang

Fig. 1   Bamboo is cut, split, and soaked in lime water, 1996
(photograph by the author)

Fig. 2   Wooden steamer (*huangguo*), 1996
(photograph by the author)

Fig. 3 Filling the steamer with soaked bamboo, 1996
(photograph by the author)

Fig. 4 Washing the steamed bamboo fibers, 1996
(photograph by the author)

Fig. 5   Modern pressure steamer, 2001 (photograph by the author)

Fig. 6   Pulping with a tilt-hammer mill (*duiwo*), 1996
(photograph by the author)

Fig. 7 Diesel or electric-powered fodder cutters have replaced
the old hammer mills, 1996 (photograph by the author)

Fig. 8 Vatman molding paper, 1996
(photograph by the author)

Fig. 9   Couching a fresh sheet of paper, 1996
(photograph by the author)

Fig. 10   Brusher pastes paper on a drying wall, 1996
(photograph by the author)

Fig. 11   Finishing paper in a *duifang* shop, 1998
(photograph by the author)

Fig. 12 The Jiadangqiao stele, 2001
(photograph by the author)

Women's work is hardly less punishing. Women in the area are rarely taller than 1.6 m, and handle sheets that are almost as long as they are tall. When brushing the upper portion of the sheet, they have to rise on their toes and stretch as far up as they can, blowing the sheet against the wall with a strong gust of breath; then they bow deep down to brush the lower part of the sheet against the wall. This movement, too, is repeated five hundred to a thousand times a day. Brushers work even longer hours than vatmen, starting around sunrise and rarely going to bed before midnight. Brushing itself takes less time than molding (the paper produced by a vatman in ten hours can be brushed in seven to eight hours), but women are less shielded from other tasks than men. Apart from short chats and cigarette breaks, vatmen rarely interrupt their work. Brushers, by contrast, are often seen shuttling back and forth between the kitchen and the drying walls, brushing a few sheets of paper in between other tasks. Mothers take their children with them to the drying shed and sometimes brush paper with their babies tied onto their backs. Women also eat less food, of a poorer quality, than men, sitting down at the table only after the men have left and even then only for a short time.

Recruitment, Training, and Control of Domestic Labor

Observers have long stressed the strong economic orientation of the Chinese family. Families were (and in the countryside still are) units of production in which senior men exercise control over the labor of juniors and women. Decisions regarding marriage and adoption, child rearing and child training, work allocation, and the like were made with the economic interests of the entire family in mind. In this system, taking a daughter-in-law or adopting a son was a method of recruiting labor, similar to hiring a long-term worker. Recruitment of male household members was rarely an issue, since men were born into a family and ideally remained there throughout their life. Families without sons could resort to agnatic adoption, known as "crossing succession lines" (*guoji*), an arrangement in which a childless man would "borrow" a nephew from a brother or same-generation cousin and raise him as his son. Adopted sons were seen as belonging to both branches, although their economic obligations lay with the families that raised them. Adoption across surname lines was less desirable, since it was seen as a transfer of loyalty and labor power for naked cash, thinly disguised in kinship terms. It seems to have been fairly common nonetheless, and the bonds between adopted sons and parents were at times very strong.[6] Families with only daughters could recruit male labor by finding a son-in-law to move in with them. Such uxorilocal or "inverse" marriages (*daocha men*) had little status and were a last resort. The position of an uxorilocally married man (*ruzhui*) was similar to that of a daughter-in-law in a standard virilocal marriage. As long as he fulfilled the obligations to his new family by working hard (typically as a vatman) and fathering children, he could not be turned out, but his status was closer to that of a hired worker than of an heir and successor. Fathers with only daughters tended to teach pulping—the task most closely identified with ownership and control—to their daughters rather than to their sons-in-law.[7]

In Jiajiang, as in other parts of China, residence tends to overlap with kinship. Many villages were dominated by a single kinship group, that is, by men who claimed common descent through the male line from the same ancestor, and by their wives and children. Such communities practiced kinship and village exogamy: sons stayed with their parents and continued the family line, and daughters married out of the village and the kinship group. Although married daughters visited their natal families at regular intervals, their principal obligations lay with their new families. For women, marriage was associated so strongly with a permanent transfer from one home to another that caterpillars and other insects were ritually "married out" to make sure they never returned:

> Buddha was born on the eighth of the fourth month.
> Today, the caterpillar is married out,
> Married out behind the blue mountains,
> Never to come back again.

or:

> Caterpillar, caterpillar,
> Black and hairy, black and hairy,

> Marry out behind the blue mountain,
> Die out, die out (*juezhong juezhong*).[8]

Most marriages in Jiajiang were "major" marriages, involving extended negotiations between two families, use of a go-between, exchange of bride price and dowry, elaborate wedding ceremonies, and virilocal residence.[9] "Minor" marriages were most common in remote mountain villages, where poor families adopted young girls and raised them as future brides for their sons (*tongyangxi*) to save the expense of a regular wedding ceremony. Another cost-reducing arrangement was to find a bride in her late teens or early twenties for a son who was much younger.[10] By doing so, families obtained the girl's labor power early on and made sure that she was given proper training. Such young brides were pitied and ridiculed in papermakers' work songs:

> A girl of eighteen, a bridegroom of nine,
> Each night she carries him to bed.
> They don't look like husband and wife,
> But like mother and son.

or:

> My husband is still small,
> Each morning I help him dress,
> Each night I carry him to bed.
> I hope that bitch of a matchmaker (*gui'erzi meipo*)
> Will die the death of a thousand cuts.[11]

Historians of Europe have argued that the growth of rural proto-industries in Western Europe was often accompanied by changes in sexual morality and reproductive strategies.[12] Under the agrarian *ancien régime*, fathers used their control over scarce land resources to shape the marriage decisions of sons and daughters. With the rise of household-based industries, it became possible for young couples to start married life early, even without their parents' consent. Moreover, since workshops could expand and thus provide additional employment for offspring in ways that farms could not, there was an incentive for young couples to have more children and have them earlier in life. These changes were accompanied by changes in sexual morality, exemplified by the spinning bees that gave young men and women opportunities to meet, chat, and flirt.[13] The lack of data on family size, marriage age, and fertility makes it impossible to tell if similar changes took place in Jiajiang, although the uncommonly small size of Jiajiang families—4.92 on average in 1944, compared to 5.92 for Sichuan as a whole—may indicate that family division took place relatively early, perhaps as a result of weakened parental control in the paper districts. What is clear, however, is that young men at least (there is no evidence on young women) dreamed of evading parental control. Romantic and sexual love is a common theme in the work songs (*zhuma haozi*) sung by men to alleviate boredom and fatigue. As often with such material, it is impossible to tell whether they depict reality or wishful thinking:

> If you plant flowers, plant *yueyue hong* (a red flower).
> If you plant trees, plant *wannian song* (a pine tree).
> If you love a girl, love one with a mind of her own,
> A mind and a will, then there'll be a way.

or:

> Rain is falling softly,
> I await my love at the middle of the mountain.
> I know my love,
> Her face is round, her eyebrows fine.
> Beauty lies in her face (*haokan zai lianpan shang*),
> Pleasure lies "halfway up the mountain" (*haoshua
>   zai banshanyao*).[14]

Training in the paper industry involved little formal instruction. Boys learned to mold, and girls to brush, by molding and brushing—playfully when they were young, seriously from their mid-teens on. Papermakers assume that skill comes easily to local people; the problem lies in acquiring the self-discipline good workers need. Boys, in particular, are literally "broken in." One papermaker told me how his sixteen-year-old son ran away three times; three times he returned to a beating, until he finally accepted his lot and became, according to the father, a good vatman. The father did not seem to mind his son's stubbornness; he even seemed to approve of it, as if the repeated beatings ensured that the skills so painfully acquired would be permanently embodied.[15] Daughters are trained in the same informal way as sons, although rarely beaten. More important than the training of daughters is that of daughters-in-law: daughters typically marry out a few years after they learn how to brush, whereas daughters-in-law stay permanently. Daughters-in-law from the paper districts may have learned the requisite skills at home, but those from the agricultural plains need to be taught to brush paper. This, too, is seen as unproblematic: as one elderly woman put it, "girls with clever hands learn to brush in a week or two; those with stupid hands take about a month."

Today as in the past, all household members work under the supervision of the workshop owner and his wife. Just as work is gendered, so is the space in which people work. Men leave the house and courtyard in the morning to work in the molding sheds, usually a short distance from home. Houses and courtyards are female spheres, reserved for women during most of the day. However, not all of women's work takes place in the privacy of the house or courtyard. Exterior walls are also used for drying paper, and there women work in full view of passersby. Moreover, drying walls are expensive to maintain and therefore constantly in short supply. Workshops with temporarily unused drying walls rent them to neighbors, in exchange for a small fee or in the expectation of reciprocity. In consequence, brushers often find themselves working in other people's living rooms and bedrooms. Encounters with unrelated men are difficult to avoid in such circumstances. Until recently, such encounters were governed by strict rules: young women had to leave the room when men entered, could not address them directly, and never sat at a table with unrelated men. Such rules protected young women from harassment but limited their mobility.[16] Footbinding, common until the 1920s, does not seem to have stopped women from working outside the household, although it undoubtedly made their work harder and more painful. Women with bound feet would use stools to support their weight during the long hours of brushing. Bound feet or not, women were on their feet for a considerable part of each day and carried heavy loads.

Recruitment, Training, and Control of Wage Labor

Most people working in the paper industry learned their skills from their parents. Formal apprenticeships were used only for boys, and only if no family member possessed the requisite skills. In such cases, a father would approach an experienced vatman, present him with meat and wine, and ask him to take care of the boy. If accepted, the boy would "do obeisance to a master" (*bai shi*) by performing a kowtow (*ketou*) in front of a scroll inscribed with the name of Cai Lun, the trade god of papermakers, and in front of his master. At the end of a three-year apprenticeship, the apprentice gave a full set of clothes to his master and received in exchange a set of tools. Apprenticeship entailed a lifelong obligation to obey the master; if a former apprentice failed to show sufficient respect for his master, the master could remove the "water nose" (*shui bizi*, a strip of wood that locked the screen in place) from the molding frame and thus stop him from molding. If the culprit replaced the water nose on his own accord, the master had the right to scold or beat him, and he was not allowed to talk back.[17]

Labor contracts in the paper industry were oral and ruled by customary norms. Most contracts began in the first month of the lunar year and were renegotiated at the end of the year; in case of conflict, they could be terminated from one day to the next. Workers and employers used kinship terms to address one another and ate at the same table, with the best and most nutritious food reserved for the workshop owner and the vatmen. Piece-rate wages ensured that workers worked fast and for long hours, but also encouraged shoddy work. Damaged sheets were sorted out and returned to the pulping vat before the paper left the workshop, and workers were not paid for such work. Nonetheless, workers could and did shirk, for example, by molding relatively thick sheets, which were easier to mold but wasted pulp. Supervision was costly, and employers knew that it was easier and cheaper to rely on workers' goodwill than to watch over them constantly. This was particularly true at times of industry expansion, when dissatisfied workers could easily find employment elsewhere.

The customary rate for male, unskilled short-term workers (*duangong*) in the 1930s and 1940s was around 1.5 kg of hulled rice per workday or the equivalent in cash; women were paid one-half that rate.[18] Wages for skilled labor were substantially higher: a vatman could earn between 2.5 and 5.0 kg of rice per day. Brushers earned about half the rate of vatmen, about as much as unskilled men. A 1923 source even states that women were paid substantially *more* than men (two to three silver *yuan* per month, against one to two *yuan* for men); this may reflect a shortage of female labor.[19] Wage levels in the paper industry were generally high: old papermakers remember that the wages of an unskilled worker were sufficient to feed "two and a half persons but not three."[20] A journal report from 1936 stated that:

Because of the abundance of Jiajiang's products, people live in peace and prosperity. Customs are pure and simple; people are used to hard work and a frugal life. Women work especially hard; in addition to fieldwork, they spin and weave. Men, besides farming, hire out to papermakers. Their daily wage and board (*gongshi*) is 7,000–8,000 cash [i.e., 0.5 *yuan*]. . . . Housing is very cheap; for 20 *yuan* a year one can rent a house of six or seven rooms. Food, housing, and clothes all are extremely cheap.[21]

According to the same source, a daily wage of 7,000–8,000 cash bought half a *dou*, or nine *jin*, of roasted rice (*huomi*)—enough to feed a family of four for two days! Wage levels in the paper industry were clearly above subsistence levels. Wage workers were not rich: they wore clothes made from handmade *tubu* cloth, patched over and over, and had little furniture apart from a bed and a chest. But they enjoyed relatively secure incomes and ate rice rather than coarse grain.[22] Most important, being a hired worker did not carry with it "the dreaded prospect of the extinction of the family line," as it did in parts of North China.[23] In the Jiajiang paper districts, no man remained unmarried simply because he was poor. In the view of one informant, only men who spent all their money on opium (*yanke*), gambling (*duke*), or prostitutes (*biaoke*) could not afford to get married.[24]

Labor Exchange and Mutual Aid

In addition to hired and domestic labor, paper workshops relied on the labor of neighbors and relatives, recruited through formal or informal exchange. Informal exchange typically involved women, who spent many unpaid hours "helping out" with urgent tasks. Such work, still very common in paper workshops, tends to go unnoticed: it is not unusual for workshop owners to claim that they use only domestic labor, even as a female relative sits in a corner of the courtyard, patiently smoothing paper sheets or cleaning *liaozi*. Formalized labor exchange was most common in seasonal steamer work, which required massive labor inputs—200 to 300 workdays—in the course of three to four weeks. Most of this labor was unskilled: carrying raw materials, filling the steamer, and pounding and rinsing the freshly steamed fibers required strength but little skill.

Not every workshop had a steamer. In fact, since only the very largest workshops steamed *liaozi* more than twice a year, and since each run of the steamer took no longer than three weeks (including loading, unloading, and rinsing the fibers), very few workshops needed to have a steamer for themselves. In the early 1950s, the ratio of steamers to paper workshops was about one to eight.[25] Most of these steamers were owned by relatively wealthy papermakers, although they could also be owned collectively by groups of smaller workshops. When owners did not need their steamers, they allowed neighbors to use them for a small fee or for free.[26] Workshops using the same steamer often engaged in labor exchange, inviting the able-bodied members of all the other households in the "steamer group" to help with heavy, unskilled tasks. Each household would request such help only once or twice per year, for a total of perhaps ten days; but since each household received help from (on average) seven other households, it needed to repay its debts at least seven times. Consequently, people spent several weeks each year working in large groups outside the household. Work around the steamer was a welcome break from the monotony of workshop work. Papermakers describe steaming as *renao*—noisy and cheerful, accompanied by song, laughter, and good food. Although exchange labor was not paid, workers expected at least one dish of "white meat" (pork belly) and one of bean curd (*doufu*) with each of the three daily meals. Hosts who served lean food were ridiculed in work songs (*haozi*):

> What kind of meat did we eat today?
> No grease on our chopsticks tips!
> You give us turnips for pork,
> Don't you fear your wealth will leave you?

No banquet was complete without large quantities of strong "white liquor" (*baijiu*, distilled from maize or sorghum)—meant, as were the white meat and white *doufu*, to ensure that the paper turned out white. Hosts who served poor liquor risked even worse contempt than those that skimped on food:

> What kind of liquor did we drink today?
> It burns in the throat and gives us a headache!
> Du Kang [the patron saint of brewers] must have
>   tipped over the still,
> Making the liquor turn sour, smelling like piss![27]

In times of crisis, households could agree to lower standards: "Let's be frugal this time; I supply only simple food, and you do the same when it is your turn."[28] However, nobody could unilaterally re-duce standards without risking a loss of social standing. Steaming was thus more than just another part of the production process; it was part of a web of exchanges and obligations that underpinned life and work in the paper districts.

Openness and Secrecy in the Paper Workshops

Stories about secrecy and concealment abound in Jiajiang papermaking, but there are few obvious secrets in the industry. The basic processes of steaming, pulping, molding, and brushing are known to all, and no effort is made to conceal them. In fact, con-cealment is impossible, since workshops are open structures without walls (Figs. 8, 9). Secrecy is also discouraged by the mutual depen-dence of household workshops that are for the most part too small to be self-sufficient. Today, as in the past, workshops swap drying walls, exchange labor, pool raw materials, borrow tools, and cooperate with their neighbors in myriad other ways. As people and equipment move back and forth between workshops, information follows. As we shall see in Chapter 7, some papermakers began to build walls around their workshops in the 1990s, but in doing so, they risked cutting themselves out of a web of mutual obligation on which they still depend.

If secrets can be said to exist at all, it is in pulp preparation. In contrast to the experiential, nonverbal skills of brushers and vatmen, pulp preparation requires a kind of knowledge best described as "folk chemistry." Pulpers use bleach for whitening the paper, dyestuffs for coloring it, sizing to make it less absorbent, and resin to make it glossy. They use hemp or bark fibers to give the paper extra "bone" (strength and texture) and straw or bamboo fibers to add "flesh" (softness and volume). Steaming is equally crucial, because the composition of the boiling *liaozi* determines the acidity of the paper and thus the way it ages. People in Jiajiang talk about this type of knowledge in terms of "recipes" (*peifang*), which they say are the exclusive property of workshop owners, who can conceal or share them at will. There are, however, reasons to doubt this claim: even though such knowledge can be verbally expressed, it usually remains tacit. Explicit knowledge, in the form of fixed guidelines or formulas, would be of little use in an industry in which raw materials are uneven in quality and too bulky to be weighed or measured. Knowledge is context-dependent: papermakers know what the pulp should look, smell, and feel like at any given stage. If some pulpers obtain better results than others, it is because they know intuitively how to respond to subtle variations in the production process, not because they possess superior fixed formulas.

This is not to say that concealment is never tried. The first person to use chlorine bleach in the 1920s bleached his pulp at night in a locked storage room and told his neighbors that he had learned a magic spell that turned the pulp white overnight. People soon found him out, and after a few weeks, chlorine bleach spread throughout the paper districts. Similarly, when a Jiajiang papermaker discovered that paper could be made thick and spongy by adding common detergent to the pulp, he claimed to have discovered a secret formula—but this, too, soon entered the common repertoire. The lack of secrecy within papermaking communities is demonstrated by the speed with which innovations spread through the industry. During World War II, when the transfer of the national government to Sichuan gave a boost to the industry, Jiajiang papermakers quickly learned how to make paper that could be used in rotating printing presses—a considerable feat, since it required changes in the format, strength, and chemical composition of the paper. The 1980s and 1990s saw a similar spurt in innovation, when household workshops introduced new technologies that reduced turnover time and labor input. Openness *in* the paper districts contrasts with the difficulty of transplanting skills to areas *beyond* the paper districts. As we will see, this is not because papermakers actively conceal their skills but because skills are socially embedded and difficult to reproduce outside their social context.

Gender and Agnatic Control over Technology

Chinese kinship, which recognizes obligations almost exclusively on the paternal side, lends itself to the creation of sharply bounded groups of agnates. In Jiajiang, as in other parts of rural China, such kinship groups become "natural" containers for technical knowledge.[29] Men are born into groups in which kinship overlaps with residence and occupation. Training is unproblematic, a function of being born into a kinship group and growing up with relatives who share the same skills. Skill is naturalized and comes to be seen as the inborn property of all men belonging to that kinship group. The advantage of this system is that there is no need for separate institutions—guilds or systems of apprenticeship—that regulate access to training. Knowledge is transmitted within existing hierarchies of gender, age, and generation; recruitment is by birth; there is no need for active exclusion or concealment because skill acquisition requires residence, and residence requires membership in the kinship group. Acute problems arise only in the case of women, who enter the community as brides but retain links to their natal families. Women's work is therefore hedged about with taboos and restrictions that discourage females from becoming competent and self-confident practitioners of the craft.

Although there is no difference in the complexity of male and female tasks in papermaking, there are fundamental differences in the way skills are embodied. A man who learns to mold can expect to work as a vatman for most of his life. Women, by contrast, are expected to leave their natal home when they marry, unlearn their old skills, and learn whatever new skills are required in their new home. Women from papermaking families who marry into the plains may have to learn to raise silkworms; women from the plains who marry into the hills have to retrain as brushers. Women are not encouraged to take open pride in their skills, and they often deny being skilled. At most, they may pride themselves on their capacity to quickly learn and unlearn a new task, rather than on competence in any given task.

This tentative, conditional embodiment of skill is partly caused by customs that identify certain labor processes as male domains and make it difficult for women to develop a proprietary attitude toward them. This is particularly true for steaming. Since the chemical processes that turned soaked, putrid, blackened bamboo stalks into white, clean, and fluffy *liaozi* were not well understood, steaming was surrounded with taboos. Men who worked at the steamers were supposed to "keep their bodies clean" (*shenshang yao ganjing*) from the polluting influence of women, that is, to refrain from sexual intercourse during the steaming period. At the steamer, men and women worked in groups, rhythmically and at great speed, to wash the dissolved lignin out of the fibers before they cooled and hardened. At one point in this process, eight to ten men, armed with long pestles, climbed on top of the steamer and pounded the steaming mass to the rhythm of improvised work songs (*zhuma haozi*). The loosened fibers were then handed down to women at the foot of the steamer, who rinsed them in nearby streams. As if in compensation for their hard work, men on the steamer were given license to ridicule and abuse any woman who caught their attention in their songs. Women who were thus mocked were not supposed to answer back; all they could do was to quickly get out of the men's sight.[30] Such ritual humiliation did not prevent women from carrying out their work, but it must have made it difficult for them to develop a sense of ownership and control over their skills.

The Locations of Skill in Jiajiang Papermaking

Skill, I argued in the Introduction, cannot be reduced to the property of single individuals or groups. Although each identifiable skill must be stored in a brain and body, skill (like other cognitive activity) extends "beyond skull and skin" into physical, social, and symbolic environments. The proper location of skill is the total "field of relations" that is reconstituted every time a skilled person involves him- or herself with a structured environment. In the case of Jiajiang, this field includes the practitioners themselves, their tools, workshops, and machines. It also includes the manmade or natural environment: bamboo groves that are not planted but shaped and maintained by regular cutting, streams that are embanked near the workshops and diverted into soaking ponds and vats. Finally, it includes social institutions and ideas that enable people to work together and reproduce their skills.

At the most fundamental level, skill is stored in the limbs and senses of practitioners. Papermaking is embodied in a very literal sense, as it transforms and deforms people's bodies. Hands and feet swell, skin grows thick and cracks under the impact of water and caustic soda. Men who still use foot-operated hammer mills, like Wang Haoding (Fig. 6), have one visibly enlarged thigh as a result of stepping up and down on a lever for several hours a day. Skills in the paper workshops are best understood as an "education of attention": skilled practitioners have learned to continually adjust their movements in response to an ongoing monitoring of the emergent task.[31] Vatmen, for example, keep their eyes on the molding frame to monitor its movement through the vat; at the same time, they respond to changing pressures in their backs, arms, and wrists as they trail the frame through the water and then adjust the speed of movement or the angle at which they hold the frame in response to this sensory feedback. Similarly, brushers feel how the soft paper sheets stretch under the brush and correspondingly adjust the pressure. Workers also use their sense of sound and smell: when squeezing excess water out of a pile of freshly formed paper, they listen for a gargling sound that indicates that water is trapped between the sheets; when mixing pulp, they smell if the pulp is fermenting and add chemicals to prevent clotting in the vat.

Papermaking skills, like other practical skills, require a structured environment for their execution. We may know how to type, knit, drive a car, or ride a bicycle, but these skills are difficult to describe or visualize until we sit down behind the steering wheel or pick up the knitting needles. Moreover, many skills are socially distributed in such a way that they cannot be carried out by a single person. It takes two to tango (or to cut a log with a two-man saw), and the requisite skills do not fully exist until the moment when two partners face and give each other a visual cue—a look or nod—to begin. In papermaking, such distributed skills are most evident in the production of large six- or eight-*chi* paper (97 × 180 or 124 × 248 cm), molded by teams of two to four workers who synchronize their movements in a slow, rhythmic dance. Socially distributed skills also exist in the continuous operations around the vat, where teams of pulpers, vatmen, and brushers work in close cooperation, and in seasonal steaming, where groups of labor-exchanging neighbors constitute a second site of skill reproduction.

To sum up, skills in the paper industry were (and still are) embodied in the brains and bodies of practitioners, situated in natural and manmade environments, and distributed across groups of practitioners. At the same time, skills are embedded in social relationships. In a sense, skills *are* social relationships, since so much of the daily social life in the paper districts—from family life to relations among neighbors—revolves around skill and its reproduction. Yet it would be naïve to assume that skill is the unproblematic property of a community, evenly shared and accessible to all. As the marginalization of women in the steaming process shows, skill is distributed in ways that reflect the distribution of power across axes of gender, generation, and class.

2　Community and Kinship in the Jiajiang Hills

JIAJIANG COUNTY LIES ON the edge of the Chengdu plain, 150 km south of Chengdu. It is linked to three different cultural and economic regions: the Lower Sichuan Basin, the Chengdu Plain, and the West Sichuan borderland (see Map 1). The county's main orientation has always been toward the Yangzi port of Leshan (called Jiading until 1913), to which it is connected through the Qingyi River, and from there to Yibin, Luzhou, Chongqing, and other cities of the Lower Sichuan Basin. Land routes and the Min River, navigable in the summer months, link Jiajiang to Chengdu, the political and cultural center of the province. Land routes also connect Jiajiang to Ya'an, the gateway to the Tibetan borderlands, and to the Greater Liangshan Mountains with their Yi-speaking populations. The Qingyi River divides Jiajiang county into two halves of roughly equal size: Hedong, which contains fertile plains and bamboo-covered hills, and Hexi, an area of high mountains (up to 1,463 m) and dense forests (see Map 2). Papermaking was concentrated in the western parts of Hedong and in Hexi, in areas where steep hills and poor soils discouraged agriculture and abundant rainfall and cool temperatures provided ideal conditions for the various types of bamboo that were the main raw material for papermaking.

In the early twentieth century, western Jiajiang was a wild frontier zone, despite its close commercial ties with Chengdu and the Yangzi ports. Hexi belongs to the same chain of mountains as Mount Emei (3,098 m), one of China's holy Buddhist mountains, and forests covered almost the entire distance from Emei to Jiajiang. The

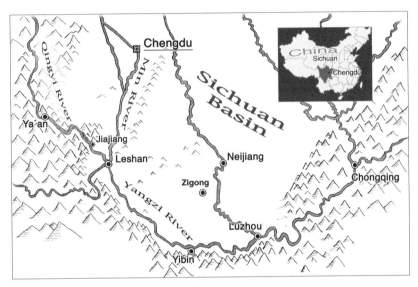

Map 1   The Sichuan Basin

county gazetteer reports that in 1897 "wild leopards appeared in the northwestern mountains and attacked people in broad daylight, so that one had to travel in groups to be safe. Magistrate He ordered them to be driven off and offered incense to the mountain god. Only then did they disappear."[1] Despite attempts to exterminate them in the early 1950s, leopards kept crossing over from Emei until 1964.[2] Black bears were sighted for the last time in 1949, right in front of the gates of the county seat. Monkeys, deer, wild boar, and wild cats were common sights until deforestation in the 1960s destroyed their habitats. In social terms, too, western Jiajiang was a frontier zone, characterized by fluid social groupings, weak local elites, an almost total absence of state power outside the county seat, and high levels of violence.

Although similar in soil and climate, the Hedong and Hexi paper districts differed in important ways. In Hedong, proximity to the county seat and to the thriving market town of Mucheng encouraged the development of relatively large workshops that used wage labor and specialized in large-format, high-quality paper. At the same time, the shortage of arable land enforced a high degree of specialization. In 1940, the average landholding in Macun township, in

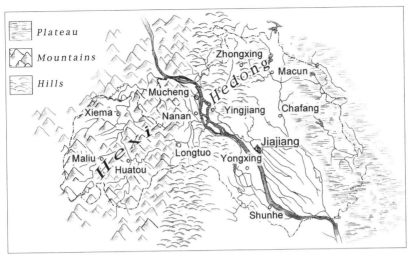

Map 2   Jiajiang: Hedong and Hexi paper districts

the heart of the Hedong paper district, was 1.73 *mu* of unirrigated farm land—too little to feed a single person, let alone an entire family.[3] In Hexi, by contrast, greater distance from the commercial centers, lower population densities, and the availability of profitable extractive industries combined to discourage full specialization. Although Hexi was more mountainous than the Hedong paper district, more of its land—both on flat mountain tops and in wide river valleys—was level; in contrast to the Hedong paper district, Hexi also contained stretches of irrigated land. Few farms in Hexi were fully self-sufficient, but with landholdings of five to fifteen *mu*, most households could feed themselves for about half of the year. In addition, the forests of Hexi provided coal, charcoal, construction timber, coffin planks, sandalwood, bamboo stalks, bamboo shoots, bamboo (which was sold for construction, or split and dried as raw material for papermaking), tea, white wax (the fatty excretion of an aphid-like insect that feeds on the twigs of a cultivated shrub), medicinal herbs, *tong* oil, and honey, as well as many kinds of nuts and fruits, both wild and cultivated.[4] Rather than investing exclusively in papermaking, households in Hexi combined agriculture, papermaking, and other sidelines in proportions that varied with market demand. Production technology in the two districts was similar, but Hedong workshops tended to invest more in their workshops and produce higher-quality writing, printing, and decorative paper, whereas Hexi workshops tended to invest little, use only household labor, and produce mainly cheap sacrificial paper (*mingzhi*) for burnt offerings to the dead.

Settlement

Like other parts of Sichuan, Jiajiang was depopulated in the wars of the Ming-Qing transition. Between 1645 and 1647, the county seat changed hands four times, being occupied in quick succession by Zhang Xianzhong's rebel army, Ming loyalist forces, and Qing invasion armies. Famine and cannibalism were widespread in these years. The county apparently recovered somewhat in the first decades of Qing rule: in 1670, a visiting Qing official noted that "when one crosses into Jiajiang, one sees paths and irrigation ditches laid out in a chessboard pattern. Smoke rises from the village hearths; it is like a scene from the Suzhou region."[5] But only four years later, Jiajiang was devastated by anti-Qing rebels loyal to Wu Sangui. It took six more years of fighting until the city was finally recaptured by the Qing.[6] Like most of Sichuan, Jiajiang was repopulated in the late seventeenth and early eighteenth centuries, largely by migrants from Hubei and Hunan. Initially, most migrants settled in the plains, but as the plains filled up, more and more settlers moved into the hills and mountains, where they produced cash crops and industrial goods for Sichuan's rapidly expanding cities.

The first relatively reliable Qing-period census, in 1812, puts Jiajiang's population at 199,172.[7] Reported land per capita was 0.56 *mu*, one of the lowest ratios in Sichuan—lower even than in the urbanized metropolitan counties of Chengdu and Huayang and in the salt-mining counties of Weiyuan and Rongxian. The low land ratio may be due in part to underreporting, common in inaccessible borderlands, but it probably also reflects a large nonagricultural population. Dependence on imported grain, brought in from neighboring Meishan and Leshan counties, was certainly a constant in Jiajiang even at that time. No township-level counts exist, but the large number of elaborate tombs, ancestral halls, and temples in the papermaking western townships indicates that they were as prosperous as the agricultural plains. The nineteenth century, often seen as a period of decline for China, witnessed relative prosperity: "The Daoguang through Guangxu reigns [1821–1908] were a period of peace when people lived without worries, and although we cannot say that 'doors were not locked,' local people lived their lives in peace."[8] In 1953, when the first township-level population data become available, the paper districts of Jiajiang were almost as densely populated as the plains.

The vast majority of the settlers in Sichuan claimed descent from Hubei, in particular from Xiaogan township, Macheng county. Such claims must be understood in the context of customary rights to land, which were acquired through the performance of specific "acts of settlement": opening the land, building houses, and burying one's dead on the land.[9] Settlement rights required the demonstration (or at least the successful assertion) that one's ancestors were indeed the first to take possession of a given place. Since Xiaogan had served twice—during the Yuan-Ming and the Ming-Qing transition—as a collection point for forcibly relocated settlers, to claim descent from Xiaogan was tantamount to claiming arrival with the first wave of settlers.[10] In the case of the Macun Shis, whose story occupies much of this and the following chapters, settlement took place in two stages: an initial landtaking in the fertile plains of Leshan, followed by a move to the hills of Jiajiang. Here is how their genealogy describes the process:

In the Wanli period [1572–1620] of the Ming dynasty, our first ancestors left Xiaogan township in Macheng county, Hubei province, and came to Sichuan. Our [founding] father died while entering Sichuan, and our [founding] mother and her three sons suffered much hardship until they arrived in Mianzhupu, Leshan county, where they settled and bought land. In order to avoid the evil of conscription, she changed the surnames of her sons to Wang, Feng, and Shi. They built a lineage hall, wrote a genealogy, and made a rule that the three surnames would never intermarry. Later, in the fifth year of the Kangxi reign [1666], our Shi ancestors, the brothers Xian, Xue, and Cai, left Mianzhupu, learned the art of papermaking, and moved to the place now known as Ancestor House Mountain. Here, they bought forest land, cut reeds to build dwellings, opened wasteland for tea cultivation, and started to make paper.[11]

The details of the story—hardship on the way from Xiaogan, death of the father, fear of conscription—suggest that the journey did not take place in the prosperous Wanli years but during the tumultuous Ming-Qing transition.[12] And while fear of conscription is a plausible motive for the surname change (since only sons were exempt from conscription), it appears more likely that the Wangs, Fengs, and Shis were not in fact brothers but unrelated migrant men who banded together for mutual protection, using fictive kinship to cement their bond.[13] A case from Huatou in Jiajiang illustrates how this could be done:

The Xiao and Luo surnames, originally from Huguang [i.e., Hubei or Hunan], and the Guo surname, originally from Jiangxi, all moved to Sichuan to settle here. The three surnames decided to build a market street [as a long-term investment] and swore an oath: "We dare not hope to always live together but vow to die the same day. Although we will enjoy the intimacy of close kin, there can be no discussion of intermarriage."[14]

From sworn brotherhood, reinforced by a ban on intermarriage, it was a small step to claiming full brotherhood and shared descent from a common ancestor—who, in the case of the Wangs, Fengs, and Shis, remained conveniently nameless.

Whatever the exact nature of the link between the Shis and their cousins, it did not prevent the Shi brothers Xian, Xue, and Cai from leaving the rich (and still sparsely populated) plains of Mianzhupu and moving to the less hospitable uplands of Jiajiang where they "bought forest land, cut reeds to build dwellings, opened wasteland for tea cultivation, and started to make paper." This suggests that the Shis formed part of the wave of "shed people" (pengmin), who, from the seventeenth to nineteenth century, opened up the hills and mountains of southern and central China.[15] These pioneer settlers, named after the simple reed shacks in which they lived, combined slash-and-burn agriculture with production of cash crops such as indigo, ramie, and tobacco, the extraction of mountain resources (coal, timber, charcoal), and industrial production.

Sichuan is unique in China for having few nucleated villages. Most people live in isolated farmsteads or in small clusters of farmhouses, scattered among the fields or lined up along mountain streams. In the paper districts, such clusters are typically built on hill slopes, so as not to occupy scarce farmland on valley floors. Some of the older clusters are built on top of steep cliffs, accessible only by winding paths that can be barricaded in case of attacks. Inside such clusters form warrens of interlocking courtyards and narrow passageways. Beyond these small clusters, whose residents are often close relatives, it is difficult to identify any territorially based communities. As Gregory Ruf has shown in his study of Baimapu village (a one-hour bus ride from my fieldwork site in Macun), village communities in this part of China are products of recent political changes, in particular of community-building efforts after 1949 that concentrated power and resources at the level of the "brigade" (dadui) or village (cun).[16] This is not to say that rural people in early twentieth-century Sichuan lived in an unstructured social landscape. Like the inhabitants of Baimapu, most inhabitants of western Jiajiang lived in a "vernacularly named landscape" in which contiguous tracts of land were marked through toponyms, graves, and buildings as belonging to certain kinship groups. The Shis of Shiyan, for example, lived in fifteen discrete settlements, many of which bore their name: Shi Weir, Shi Cellar, Shi Shop, and so on.[17] "Shi territory" bordered on other areas of contiguous single-surname settlement: Yangbian (inhabited by the Yangs), Zhangyan (by the Zhangs), Chadigeng (by the Lis), and so on.

Despite active land markets in the area, it was difficult for outsiders to acquire property. A 1941 land survey of the ninth bao of Macun (which covers much of today's Shiyan village) shows extreme fragmentation of landholdings, with some 1,400 plots held by 138 owners and an average plot size of 0.42 mu for arable and 5.92 mu for forest land. Papermaker Shi Zhixuan, for example, owned four buildings (zhai), one-third of a molding vat, one-quarter of a pulp mill, eight plots of dry farmland, two plots of forest land, and one plot of bamboo land, but these seventeen pieces of property amounted to only 11.5 mu (1.9 acres) of land.[18] As elsewhere in China, local customs required that land for sale was first offered to relatives and neighbors, and there was a strong aversion—expressed in the common saying that "fertilizer should not be allowed to run off to other people's fields" (feishui buyao wailiu)—to selling land to outsiders. Only twelve of the 138 property owners listed in the survey were non-Shis: four Wangs, three Yangs (presumably belonging to the neighboring Yang family of Yangbian), and one member each of the Li, Zeng, Deng, Huang, and Xue surnames. Such "outside surnames" (*waixing, yixing*) often started their life in the area as dependent clients of individual Shis. Papermaker Xiong Yuqing, for example, recounts how his father, a stone mason, came to Shiyan in the 1930s. At a time when common people "looked upon conscription like death," he fell into the hands of a press gang and nearly perished from malnutrition and abuse in the camp where he was held.[19] He escaped by crawling through a sewer, fled to the hills, and hid in the house of the influential papermaker Shi Ziqing, who at that time needed a mason. When the commander of his unit came to demand him back, Ziqing bribed or intimidated him into letting Xiong go.[20] Xiong later married a Shi woman; his son Yuqing learned to make paper, became a successful paper trader, and served for a time as Shiyan's village head. Although outsiders could become fully integrated members of the village community, their acceptance depended on the goodwill and support of influential Shis.

The *Baojia* System

The nearest equivalent to the village in the paper districts was the *bao*, a numerical unit that in theory comprised ten *jia*, each of which in turn consisted of ten households. Such numerical units had a long tradition in China and were reintroduced by the Republican government in the 1930s. In the paper districts, a *jia* typically corresponded to a small settlement core, whereas a *bao* might include ten to fifteen settlements and about one to two hundred households. Each *bao* had a headman (*baozhang*) and deputy (*fu-baozhang*), appointed by township officials, who apportioned taxes and decided who would be drafted into military service. These positions carried real power, especially during the anti-Japanese war, when the threat of conscription could be used to intimidate villagers, but they conferred little prestige and were shunned by wealthy and influential men. The *bao* had little social content: it was a unit of registration, taxation, and control, not a community of shared interests and sentiments.

The *Paoge*

Kinship, discussed below, was the most important principle of social organization in the paper districts. Most people in the hills spent their days surrounded by agnatic kin and met non-kin only during trips to the market town. However, people who aspired to more than average power and prestige needed to interact regularly with non-kin. One of the most important channels for such interactions was the *paoge*, or "society of robed brothers," also known as the *gelaohui*, "society of elders and brothers." The roots of this organization lay in mutual help networks among smugglers and transportation workers and in the anti-Manchu agitation of the early twentieth century. By the 1920s, it had transformed itself into a multipurpose network without a strong ideology, deeply entrenched at all levels of Sichuanese society. By that time, there was nothing secret about the society: in Jiajiang, one man out of five—most officials, most military men, and most merchants—were members.[21] Later accounts often differentiate between "clear water" (*qingshui*) and "muddy water" (*hunshui*) factions, but, according to former members, most lodges engaged in both "clear" (legal) and "muddy" activities. The *paoge* were divided into five grades: *ren* (benevolence), *yi* (righteousness), *li* (propriety), *zhi* (wisdom), and *xin* (faith), which recruited members from different social strata and formed separate lodges.[22] Lodges were led by "helmsmen" (*duobazi*), locally influential men who could settle disputes, negotiate with other helmsmen, or accommodate requests from traveling brothers. This last was, in the opinion of former members, the most important function of the *paoge*: if one ran into trouble with people from outside one's own kinship group and home community, one would seek out the nearest *duobazi*, who would sort it out. This was especially important for men who traveled, including merchants, soldiers, bandits, itinerant repairmen, and beggars. Farmers and small papermakers traveled little and therefore had no need to join the organization.[23]

The *paoge* overlapped with other structures in ways that made it difficult to tell if the authority of a leader derived from his rank in the *paoge* or from his other functions. For example, Shi Longting, the overall *paoge* leader (*zong duobazi*) in the Shiyan area, was also much respected among the Shis for his high generational status and his skill in settling disputes. Longting, a successful papermaker, owned a teahouse in the heart of the area inhabited by the Shis. Commercial and personal disputes were brought to the teahouse, where Longting and other senior men (who spent much of their time in the teahouse or could be called in from nearby) would listen to both sides and propose a settlement. All teahouse customers could participate in these discussions; if they disagreed with Longting's proposal, they would hoot, shout, and bang their teacups on the table, and the debate would continue. The only sanction at Longting's disposal was to order plaintiffs or defendants to buy tea for all those present. Nonetheless, the teahouse court settled most conflicts in the area.[24]

Religious Associations

Religious associations were an important aspect of social organization in Jiajiang, not least among papermakers, who worshipped Cai Lun, the inventor of paper and protector of their trade.[25] An undated source from the last years of the Qing dynasty lists 32 Buddhist temples, eight convents (an), and 40 Daoist temples in the county. Most of these temples were closed and their assets confiscated during the "antisuperstition" drives of the late Qing and early Republic.[26] D. C. Graham, an American missionary and ethnographer who visited Jiajiang in 1945, reported:

In the city and within a radius of 10 miles were 30 temples, 4 in the city and 26 outside. Three temples in the city and two outside were reported to have been sold and destroyed; the other temple in the city was occupied by a military school. It seems a very safe guess that at least 15 former temples in the city had ceased to be temples for so long that they were not reported to us. Twelve temples outside the city were used for worship only; some of these were too small and others too far away to be used for other purposes. Several of these were in a poor state of repair. . . . Three temples were used as schools, two as waterpower plants, and one each by a youth organization, a military center, a post office, a theater, a charity organization, and a barracks for soldiers.[27]

It was only in remote Hexi that temples remained an important part of the social scene until the 1940s. Not all temples served public functions: some of the smaller convents were maintained by rich families who used them to "raise nuns" (*yang nigu*) to keep the women of their family company, entertain guests, and help mourn when a family member died.[28] Most temples, however, served entire village communities, and some of the larger temples hosted annual fairs (*miaohui*) that attracted thousands of worshippers. The most important temple fair in the paper districts was that of the Dongyue temple in Huatou. Dongyue, the God of the Eastern Peak, was believed to hold power over life and death, wealth, and honor and was widely worshipped in Sichuan.[29] The Dongyue fair in the third lunar month lasted three days, during which the streets of the town were filled with worshippers, peddlers, and food stalls. Operas were staged in the temple; people burned incense and sacrificial paper money on the streets until the ash piled up in huge heaps. The fair culminated in a procession of the god's statue through the town and the surrounding countryside, accompanied by costumed dancers and men with knives stuck in their arms to express their devotion.[30]

Such fairs were organized by "incense and lamp associations" (*xiangdenghui*) composed of wealthy and respected members of the community. The Dongyue *xiangdenghui* extended to all parts of Hexi, with branch associations in many market towns.[31] Managing the Dongyue temple fair added enormously to a man's prestige and could yield a handsome profit, but the financial risk was considerable. Several months before the fair, the leaders of the branch associations told the association head (*huishou*) how many members from their townships would attend and pledged (*yue*) a sum of money in accordance with the number of participants. However, most expenses for the theater troupes and for the banquet (up to five hundred tables—four thousand guests, if we assume the standard seating of eight persons per table) were paid by the association head out of his own pocket and recuperated later.[32]

Violence and Power

Republican-era Jiajiang was a violent place. Decades of warlord rule had led to a proliferation of firearms among the population, and people of all social strata, from vagrants to local elites, had recourse to violence to pursue their aims. Most rich people kept guns at home, and local leaders often kept armed retainers. Bandits (known as "cudgel men," *bangke* or *bang lao'er*) robbed houses and kidnapped children and women for ransom.[33] People in Shiyan alleged that many of their neighbors in Zhangyan were "farmers by day and bandits by night." Although it was said that "eagles don't hunt close to their nests," the Shis kept their distance from the Zhangs. If attacked by bandits, one had to stay flat on one's belly, remain quiet, and not look up; then one would not get hurt. The county had no police force; army garrisons, nominally in charge of maintaining law and order, were heavily involved in drug trafficking. Army commanders would insist on payment before sending out their troops, and soldiers would start shooting long before they reached the bandit "nests" (*tufei wozi*) so that the bandits had plenty of time to escape. The best protection against bandits was to buy them off, but even that was often difficult. Shi Ziqing, one of the richest papermakers in Jiajiang, went so far as to visit the bandit leader Zhang Gaishan at home and offered him "fixed employment, meals and opium twice a day, and wages whether he worked or not." Zhang refused the offer and later tried to ambush Ziqing as he was returning from a business trip. Ziqing escaped unharmed, but instead of seeking revenge he repeated his offer until Zhang finally accepted.[34]

Violence often carried over into local politics. Throughout Jiajiang, local elites used their connections with *paoge* and bandit gangs to intimidate or eliminate opponents, but the style of violence differed markedly between Hexi and Hedong. Hexi was dominated by landowning families, linked through the *paoge* to the networks of military men that ruled Sichuan until 1936. Hedong, by contrast, was dominated by commercial elites with connections to the urban centers of Chengdu and Chongqing. Two episodes serve to illustrate the difference. In 1939, He Xicheng, a *bao* headman in Hexi, invited local notables to celebrate the marriage of his son. When the men sat down for dinner, two *duobazi*, Wu Zuozhang from Huatou and Zhu Guangrong from Emei, started a violent dispute over precedence at the table. Through mediators, the men agreed to meet in Huatou for an armed showdown. Both men had links to army units: Zhu received reinforcements from the new 17th Division (under Sichuan's strongest warlord Liu Xiang), and Wu called on the 24th Army (under Liu Xiang's uncle and rival Liu Wenhui). On the agreed date, Zhu's forces rode into Huatou, where Wu's men lined the streets, cudgels in hand, guns hidden under their coats. Before fighting broke out, however, the overall leader of the Eight Great Lodges (ba da gongkou) of Huatou intervened. The leaders of the hostile camps were called to an improvised banquet and plied with liquor until they agreed to accept mediation. Zhu was found guilty and ordered to deliver 2,000 fir trees to Wu's party to recompense them for "powder and bullet costs."[35]

In Hedong, rivalries were settled no less violently but in a less ritualized way. Politics in Hedong were shaped by the rivalry between the Wang family, based in Jiajiang city, and the Jiang family in the prosperous township of Yanjiang. The Wangs' main connections were with officials appointed by Liu Xiang's warlord regime and with the paoge; by contrast, the seven Jiang brothers (whose number included a Beijing university graduate and a graduate in economics from the Sorbonne in Paris) sympathized with the Guomindang. Both sides were involved in the paper trade, but also vied for control over arms and opium smuggling. The Jiangs used "modern" institutions—schools, study societies, debating clubs—to a much greater extent than the Wangs but were not above mobilizing their paoge allies for paid assassinations. In the 1947 elections for the national assembly, both sides fielded candidates, and the "ballot elections" (piaoxuan) soon deteriorated into "bullet elections" (paoxuan). One prominent victim was Ma Weishan, one of the most powerful paoge leaders in Hedong, who was dragged from his car and shot on his way to a wedding.[36] Violence in Hedong was more "modern" not only in terms of equipment (horses and cudgels in Hexi; cars and hand grenades in Hedong) but also in the absence of chivalric codes of conduct. The financial stakes were also much larger in Hedong.

Kinship in the Paper Districts

In an area where the presence of the state was hardly felt, where there were no strong village communities and no entrenched bureaucratic elites, most people's daily lives were structured above all by kinship. People in the paper districts worked, relaxed, and socialized mainly with relatives; when they interacted routinely with members of a different kinship group, they tended to establish fictive kinship bonds so as to avoid the uneasiness of dealing with "strangers." To be sure, people mixed and mingled with nonrelatives during their visits to market towns, and men with commercial or political interests socialized with unrelated men in teahouses and restaurants. But for most Jiajiang people, these were but brief interruptions of daily routines that revolved around kinship.

Like most authors interested in Chinese kinship, I take the work of Maurice Freedman as my point of departure. My aim is not to refute Freedman's "lineage paradigm," which has been subject to numerous criticisms and qualifications from the moment of its inception,[37] but to point to a realm of practical, use-oriented kinship that overlaps with the paradigmatic lineage. The concern here is not with the *vertical* descent lines crucial to Freedman's model but with *horizontal* ties between male agnates of the same generation. At the heart of this form of kinship lies the practice of *beifen*—literally, distinction (*fen*) between generations (*bei*). This is a perfectly orthodox element of Chinese kinship, practiced in all parts of China and resonant with widely shared and deeply held moral values but almost totally ignored in Western studies of Chinese kinship.[38] My argument in a nutshell is that kinship is not primarily a system of rules and representations but something that people *do* in their daily life. For people in Jiajiang, kinship mattered to the extent that it was emotionally satisfying and helped them achieve concrete results for themselves and their family. In an environment in which most economic transactions involved relatives, one central function of kinship was to help people secure loans, hire workers, borrow equipment, and so on. *Beifen*, with its emphasis on horizontal ties between lineage "brothers," helped people to achieve these aims. Equally important, it ensured that production-related knowledge was not monopolized by single descent lines but shared among all agnates.

Following Freedman's lead, anthropologists of China have long focused on the formally constituted descent group, or lineage. The attraction of Freedman's work is easy to understand: by demonstrating how the Chinese used a number of simple but consistent principles (unilineal descent, patrilocal residence, agnatic exogamy) to construct descent groups of varying size but homologous structure, Freedman could depict Chinese society as being built from the ground up. Descent groups—families, extended families, lineage branches, sublineages, lineages, and clans—formed nested hierarchies; at each level, these hierarchies aggregated people and accumulated resources into units that competed with one another for wealth and power. At the highest level, descent groups (or their elite representatives) met and meshed with the lowest level of the state. In this way, Freedman could explain how China had achieved such a high level of social and cultural integration, despite the well-known fact that the imperial state never reached down to the village level. Organized kinship was the link between the Confucian state and grassroots society: from the top, it disseminated state-sponsored orthodox values; from the bottom, it aggregated dispersed interests and gave them social expression.[39]

Chinese kinship, in this formulation, is centrally concerned with land and power. Descent groups compete, often violently, for influence and power; land is both the prize and the precondition of power. The strength of descent groups varies in direct proportion to their control over corporate resources: only groups with large estates can afford the lavish rituals and provide the material benefits—cash or grain stipends, education in lineage schools, land rented out at below-market rates—that keep agnates together generation after generation.

Yet, as Freedman pointed out, lineage solidarity was always threatened, since the same organizational principles that facilitated lineage creation made it possible to split an existing lineage into competing segments. Much of the study of the Chinese lineage has focused on asymmetrical segmentation, that is, the process by which a subset of a larger kinship group selected an ancestor that they (but not their more distant cousins) had in common, endowed an estate in his name, and thus set themselves apart from their cousins. As Rubie Watson has shown, this was often class differentiation in disguise, as privileged lineage members set up separate estates in the name of their fathers, making sure that the benefits from these estates went only to their own descendants and not to those of poorer branch lines.[40] Lineages, it seems, were characterized by strife and competition between agnates, mitigated only when the lineage came under external threat. Corporate resources tended to be monopolized by powerful lineage members, who ran lineage estates "as hard-nosed businesses and [not as] charitable estates."[41]

Formal Kinship Organization in Jiajiang

How does kinship in Jiajiang fit into this picture? First of all, there is nothing unorthodox or unusual about it. People in Jiajiang venerated focal ancestors and sacrificed at their graves; their associations (hui) owned property and organized ritual activities during the major festivals; some larger kinship groups wrote genealogies and maintained ancestral halls. Because of their relatively recent arrival in Sichuan, lineages and other descent groups were smaller and had less genealogical depth than did similar groups in other parts of China.[42] Ancestral halls (citang) were relatively common: in a county with 183 different surnames (and an even larger number of descent groups, since some unrelated groups shared the same surname), there were 77 halls in the 1940s.[43] Corporate property tended to be small: the Shi lineage, for example, owned "as much land as a middle peasant." All this puts descent groups in Jiajiang somewhere near the middle of Freedman's "A to Z continuum" of small and simple to large and complex descent groups.

As elsewhere in Sichuan, the management of corporate property lay in the hands of Qingming associations (Qingminghui), named after the banquets they organized during the Qingming festival in the spring.[44] These associations were led by committees of well-to-do men of good standing, who selected a headman (huishou) from among their number. The Shis' Qingming association was closely associated with the Zhongshan temple, a small Buddhist institution in the hills above Macun that received most of its funding from individual Shis. The yearly Qingming banquet was held in the temple, and the head monk of the temple managed the association's land. Similar arrangements existed elsewhere in Jiajiang: the Huatou gazetteer reports that wealthy descent groups endowed Buddhist or Daoist temples (si, miao), which came to function as their "family temples" (jiamiao), although they remained open to the public.[45]

Qingming celebrations began with the cleaning and repair of ancestral graves, followed by the offering of incense and paper money and the ritual prostration of all descendants in front of the grave. The graveside sacrifice was followed by a banquet, known as "eating the Qingming association" (*chi Qingminghui*). Although participation in the graveside rituals was limited to household heads, the banquet was open to women, children, and even members of other kinship groups. The Shis expected all households, even those living far away, to send at least one member; in addition, the Shis invited neighbors and locally influential men. Elderly Shis recall, often with glee, how disobedient sons and daughters-in-law were punished during Qingming celebrations. Elders could punish mild violations of the "family code" (*jiaxun*) on the spot; more serious transgressions were brought before the Qingming association, which ordered offenders to kneel for the duration of the banquet or to contribute money to the repair of streets and bridges. In the most serious cases, offenders were made to bend over the "family regulation bench" (*jiafatai*) and beaten.[46]

Some local Qingming associations were constituents of "large" (i.e., higher-order) Qingming associations.[47] The Shis, together with their Wang and Feng cousins, maintained a "joint ancestral hall" (*zong citang*) in Mianzhupu. According to Mianzhupu residents, the hall, courtyard, and attached opera stage covered an area larger than a football field; during Qingming celebrations, the courtyard seated more than 300 guests at 40 tables.[48] Despite its name, the joint Qingming association did not convene on Qingming day but in the ninth lunar month. Every three years, it organized a "big celebration" (*da qing*), attendance at which was obligatory for affiliated families; participation in the "small celebrations" (*xiao qing*) in intervening years was voluntary. During the festivities, Wangs, Fengs, and Shis intermingled, ate at mixed tables, and jointly bowed to a stele dedicated to the "successive ancestors of the Wang, Feng, and Shi surnames" (*Wang Feng Shi shi lidai zuxian*). Like the Shi Qingming association, the joint association had a council that could punish members for unfilial behavior (*bu xiaoshun*), arrogation of generational rank (*weibei paihang*), and intermarriage between the three surnames (*luan kai qin*). Those found guilty could be scolded, fined, beaten, or forced to kneel in front of the stele for the duration of the banquet.[49] Elderly Shis also recall that before 1949, Shi men passing through Mianzhupu would visit the ancestral hall and prostrate themselves in front of the ancestral stele. Those who were able to recite the generation names (*paihang zi*) laid down by the ancestors were given food, accommodation, and enough "opium money" (*yan-qian*) to indulge themselves for three days.

Apart from the collective Qingming sacrifices, families worshipped their ancestors individually. In contrast to nearby Baimapu, where graves in lineage cemeteries were arranged in neat rows and lines, expressing generational rank and patrilineal descent, graves in the paper districts were scattered across the hills.[50] During the major festivals (New Year's, Qingming, and Zhongyuan, on the fifteenth day of the seventh lunar month), people visited the graves of ancestors they had known in their lifetime; graves of earlier ancestors were allowed to fall into disrepair. At home, people offered food and incense to close ancestors during important festivals and on the first and fifteenth of each month. Such sacrifices took place in front of a large sheet of red paper pasted on the rear wall of the central room of each home. The central column on the paper read "spirit seat of the successive generations of the enshrined male and female ancestors of the X surname" (*X shi tangshang lidai xianzu kaobi zhi shenwei*); columns to the left and right named such deities as Heaven and Earth (*tian di*), Guanyin (*Guanyin pusa*), Confucius and Guandi (*wen wu fuzi* or *Kong Guan ersheng*), the Gods of the Kitchen and the Earth (*zaoshen tudi*), and the Gods of Happiness, Office, and Wealth (*fu lu cai shen*). This list of "family spirits" (*jiashen*) often included trade gods, such as the deified inventor of paper and "First Teacher" of papermakers, Cai Lun (*Cai Lun xian shi*).[51] Wooden tablets for individual ancestors were used only by wealthy families, and only in the first three years after a person's death. After three years, the spirit was assumed to have left the tablet and moved to the grave.[52]

Generational Order

Descent groups in Jiajiang, as in many other parts of China, practiced a system of generational naming based on lists laid down by the ancestors. In the case of the Shis, the founding brothers Xian, Xue, and Cai decreed that their sons would use the character *wei* in their name (Shi Weilong, Shi Weihuan); grandsons would use the character *ke* (Shi Kefu, Shi Kesong), great-grandsons the character *xing*, and so on through twenty generations. Daughters were not given generation names since they left the descent group when they married. Although most men used their generation names in public, this was not obligatory; men with formal education often adopted "school names" (*xueming*) for themselves and sometimes acquired other names over the course of their life.[53] However, even those who did not use their generation name were clearly understood to have one. As the man known as Shi Dongzhu explained, "My parents gave me the name Shengquan. Even if I don't use it all my life, this is my real name, the one by which I'll be remembered when I'm dead. In the underworld, I'll be Shi Shengquan."

Generational naming formed part of the system of kinship ideologies and practices known as *beifen*, "distinction between generations." In descent groups too large for all members to know and recognize one another, generation names helped to identify relatives as generational seniors, juniors, or equals. For example, a man called Shi Quanyou would immediately recognize his distant cousin Shi Guijie as a member of the *gui* (or thirteenth) generation, one step above his own, the *quan*.[54] He would therefore address him as *daye*, the term appropriate for a man in his father's generation, and treat him with the respect due to a paternal uncle.[55] Since 1949, *beifen* has lost much of its force, but it still governs behavior in many social situations. Until quite recently, generational juniors were not supposed to laugh or talk loudly in the presence of their elders; when a senior person entered the room, they had to rise and remain standing until given permission to sit down. Disrespect (or a simple failure to get out of the way of an ill-tempered senior) could result in scolding or a slap in the face. Even today, generations sit at separate tables during formal banquets, and there is a tendency for men of different generations to avoid eating together, even at home. Generational equals, by contrast, were expected to behave "like brothers," in other words, not to stand on formality, to engage in rough jokes and horseplay, and to help one another. It is important to note that generations lacked intrinsic content: the members of the *tian* generation, for example, did not form a group, nor did all *tians* necessarily know one another, nor did they possess qualities that set them apart from other Shis. The generational structure with its subordination of juniors to seniors could be harshly hierarchical, but this was mitigated by two facts. First, seniority was always relative; a *gui* man was senior to a *quan* in exactly the same way as a *sheng* was senior to a *gui* and a *ding* was senior to a *sheng*. Second, high *beifen* status had nothing to do with merit. It was a matter of luck (some men were born into more senior generations than others) and of the passage of time (men who lived long enough could expect to become senior to most others). This automatic and impersonal progression blunted the edge of generational privilege.

The broad categories of generational seniors (*zhangbeizi*), juniors (*xiaobeizi*), and equals (*tongbeizi*) were used mainly in dealing with remote kin. Close relatives addressed (and still address) one another with precise kinship terms. These terms, however, tended to blur differences between relatives of the same generation. Paternal uncles, for example, were addressed not with the standard Chinese term for older or younger uncle (*bobo, shushu*) but were called "first dad," "second dad," (*daba, erba*), and so forth, the prefix indicating that they were the first, second, and so on sons of their parents; their wives were addressed as first or second "mom" (*daniang, erniang*). Differences *between* generations, in short, were constantly emphasized; differences *within* generations were downplayed to an extent that made it difficult to see if two men of the same generation were sons of the same father or distant cousins.[56] The equality and, to some extent, exchangeability of men in the same generation is also expressed in the custom of *guoji*, or agnatic adoption, discussed in the previous chapter. In the name of agnatic solidarity, men could be asked to donate a "surplus" son to an heirless cousin, and men with particular skills could be asked to foster, employ, or train the sons of unskilled cousins.[57]

In principle, generation overrode sex, age, and social status.[58] Because of uneven birth intervals over many generations, it was possible for a man in his seventies to be the generational "son" or "grandson" of a teenager. *Beifen* rules mandated that the older man address the younger with the appropriate term for a senior (*daye, laoye,* or *laogong,* depending on the precise generational distance), whereas the younger man addressed the older informally as "you" or by personal name. In practice, young but generationally senior men tend to avoid this awkward situation by using the respectful "teacher" (*laoshi*). *Beifen* is also avoided or downplayed when it clashes with social status: generationally senior workers call their employers *laoban* (boss), even though they are entitled under *beifen* rules to use personal names. In turn, the employers address them with appropriately respectful kinship terms.

In theory, *beifen* exists only between agnates, but it is easily extended to nonagnates. Wives automatically received the generational status of their husbands, and uxorilocally married men received the status of their wives (or, more precisely, that of the wife's brothers). Maternal relatives and affines could be similarly slotted into the system: a mother's brother or a wife's father could be treated as belonging to the next higher generation. In the absence of direct kinship ties, generational equivalence could be constructed on any number of grounds: men could, for example, consider each other "brothers" (*dixiong*) because their grandfathers had been friends or had gone to school together. In places with one dominant descent group, everybody else adapted to their generational system: in Shiyan, for example, almost all non-Shis are descended from Shi mothers or married to Shi wives or husbands; they consequently "speak" (*jiang*, i.e., practice) *beifen* with the same ease as Shis.

Kinship That Works

Kinship in the Jiajiang hills systematically de-emphasized lineal descent. At home and at graves, people worshipped only those relatives that had been an active presence in their lives; all others were worshipped collectively and anonymously as the "successive generations of enshrined ancestors." None of the many methods available in Chinese culture to record lineal descent—genealogies, inscriptions on stelae and tombstones, placing of graves and tablets—was used for that purpose. Tombstones and stele inscriptions recorded generation names rather than information about descent lines.[59] Even "genealogies" (*jiapu*) did not necessarily contain genealogical information: the Shis' genealogy (now lost) contained a list of the twenty generation names but no information about who begat whom. In ritual terms, kinship groups were seen as consisting of horizontally layered generations rather than of vertical lines of descent.

This "horizontal orientation" had two practical consequences. It discouraged the formation of privileged lineage branches that monopolized resources for themselves, and it provided a basis on which all members of a descent group—and indeed nonmembers who had learned to "speak" the same *beifen*—could easily establish ties for a wide variety of purposes. Branch formation was not entirely unknown in the paper districts: old Shis remember the story of Shi Wanshun who lived in the late eighteenth century, became miraculously rich, and left fields and a large house to his descendants. His grandsons, collectively known as the "Seven Golds" (because they belonged to the *jin* or "gold" generation), established a Qingming association in his name that met every year after Qingming for sacrifices and a banquet.[60] Although this separate ritual focus may have created a sense of shared identity among Wanshun's descendants, they continued to use the same *beifen* names as their cousins and thus remained part of the larger network of claims and obligations that tied all Shis together.

Crucially, technical knowledge—a more important resource for papermakers than land and other physical assets—could not be monopolized for individual descent lines. The Shis, and presumably other descent groups as well, saw the "art of papermaking" as a gift from the ancestors to all their descendants, not to particular families or descent lines.[61] It is likely that successful papermakers wanted to ensure that certain skills or knacks remained in the hands of their sons and grandsons, but such "selfish" favoritism toward one's descendants was seen as illegitimate and could be given only muted expression.

In the view of contemporary informants, the hierarchical aspects of *beifen* are less important than the fact that *beifen* relationships are characterized by a degree of closeness and emotional warmth that set them apart from relationships "in society" (*zai shehui*), that is, outside the kinship group. On one hand, *beifen* relationships are modeled on some of the most important bonds that people have in their lives: those with parents, siblings, and children. Such relationships are familiar, easy to comprehend, and emotionally satisfying. Like all kinship, *beifen* is at least partially about obligations and rights, but its broad and inclusive categories do away with the kind of cold kinship arithmetic that asks what exact claims can be made on a third paternal cousin or on a brother's wife's sister's husband. *Beifen* provides people with a language of moral and emotional commitment that covers almost all everyday situations. At the same time, *beifen* is flexible enough to be used strategically. One does not need to "speak" *beifen* all the time, or speak it with the same degree of intensity; particular genealogical links can be invoked in particular moments and filled with emotional content.

Communities of Skill

To a very large degree, the economy of the Jiajiang hills revolved around intangible assets. The industry as a whole depended on the unimpeded circulation of technical know-how, and the success of individual papermakers depended on their ability to establish and maintain a wide range of relations with buyers and sellers, neighbors and wage workers, and on their accumulation of "credit" in the broad sense of trust and reputation. Land and other fixed assets appear to have been far less central to social relations than they were in many agrarian contexts. Kinship ideologies and practices in the paper districts appear well adapted to such an economic structure. Yet we must not assume that kinship and other social structures in Jiajiang are a direct reflection of underlying economic structures. Edmund Leach's formulation that kinship is "another way of talking about property relations," though useful, is too narrow; and nothing would be gained by defining kinship as "another way of talking about technology."[62] Moreover, there is little evidence that kinship (and other dimensions of social organization) differed radically between the papermaking hills and the agricultural plains of Jiajiang. If the relatively flat and inclusive practices discussed above appear atypical, this is perhaps because my focus on kinship in the context of collaborative work throws into relief practices that are quite common but remain obscured as long as we see kinship as concerned primarily with control over property.

3   Class and Commerce

THE WEB OF OBLIGATIONS that linked papermakers to one another mitigated class differentiation but did not prevent it altogether. Inequality existed throughout the paper districts. It was most pronounced in the developed and commercialized parts of Hedong, where the industry was dominated by "big households" (*dahu*), fully specialized workshops that employed skilled workers, kept their workshop in constant operation, produced paper of high quality, and maintained mutually beneficial trade relations with large merchants. In Hexi, "small" households (*xiaohu*)—sideline producers who used mainly household labor, closed their workshops during the agricultural season, and produced paper of poor quality—were more common; wage labor was comparatively rare, and income differences were less pronounced. Inequality probably varied not only across space but also across time, but the paucity of pre-1940s data makes it difficult to ascertain this. The first half of the twentieth century saw a rapid expansion of the Jiajiang paper trade, brought about by the burgeoning economy of Sichuan's urban centers, improved transportation, and the transfer of the national government to Sichuan after the outbreak of the Sino-Japanese war. This general expansion was punctuated by sharp crises: warlord hostilities in 1933–35, a province-wide famine in 1936–37, and the collapse of Sichuan's wartime economy after 1945. Both the trade expansion in the early twentieth century and the crises in the 1930s–1940s seem to have favored large, wage-labor-employing workshops, which were better equipped to weather economic storms. In parts of Hedong, polarization between rich and poor seems to have increased in the first half of the twentieth century, although not to the point that traditional obligations broke down.

Table 2
Paper Workshops by Output, 1952
(unit: 10,000 sheets *duifang* paper)

Output	Equivalent to workdays	Households	As percentage of all households
0–1	0–230	418	8.2%
2–5	231–1,150	2,329	45.7
6–10	1,151–2,300	1,375	26.9
11–20	2,301–4,600	484	9.6
21–30	4,601–6,900	86	1.6
> 30	> 6,900	401	8.0
TOTAL		5,093	100.0%

SOURCES: Cheng Quan, "Jiajiang zhishi," 22; Sichuan Archives, Gongyeting 1951 [13], 7.

According to data from the early 1950s, 30 percent of papermakers were "sideline producers" (*fuye zaozhi*), and 42 percent were "half farmers, half papermakers" (*bannong banzhi*). Another 17 percent relied "mainly" on papermaking (*yi zhi wei zhu*), and only the top 11 percent were "full specialists" (*chun zaozhihu*). Another statistic from the early 1950s (Table 2) shows that 54 percent of papermakers produced fewer than 50,000 sheets per year. Assuming a labor input of 230 workdays per 10,000 sheets, these workshops spent up to 1,150 workdays per year on papermaking.[1] Since this is equivalent to the full-time labor of 3.5 workers, a household workshop at the upper margin of this category could be considered fully specialized. Workshops in the 50,000 to 100,000 sheets category (27 percent of the total) must have been fully specialized and very likely used some hired labor. Workshops producing over 100,000 sheets annually (19 percent) must have employed the equivalent of at least seven full-time workers, domestic or otherwise, in their workshops. These were the "big households" that produced paper all year round and relied heavily on hired labor. A rough estimate, then, is that more than half of paper workshops operated seasonally (although even these might derive more income from papermaking than from farming), one-quarter were fully specialized and used primarily domestic labor, and about one-fifth were large-scale business operations employing primarily wage labor.

Ownership structure differed widely between areas. The area inhabited by the Shis in Hedong had one of the highest concentrations of paper workshops and one of the most polarized income distributions. The 33 vats active at the time of land reform, in 1951, were owned by fifteen workshops. The biggest workshop, that of Shi Guoliang, operated seven vats; Shi Haibo had five, Shi Longyu three; thus, almost half of all vats were owned by just three workshops. Six workshops owned two vats each, and another six owned a single vat. The majority of the local population—two-thirds or more—worked for the "big households." In Hexi, the ownership structure was much flatter. Sideline producers had little need for wage labor, and the few workshops that hired workers tended to recruit them from Hedong.

How difficult was it for a worker to open a workshop, or for a small producer to join the ranks of the "big households"? In their investigation of paper workshops in Yicun, Yunnan, Fei Xiaotong and Zhang Zhiyi reported that the average workshop represented an investment of 1,000 *yuan*—a substantial sum in the early 1940s, although not more than the typical yearly income of a workshop.[2] Almost all equipment in Jiajiang was produced locally from sandstone, wood, bamboo, and other materials found in the area. The most expensive component of paper workshops in Jiajiang was the drying walls (*zhibi*), which were made by plastering the outer walls of houses with a mixture of chalk and paper pulp that was polished until it became as hard and smooth as glass. A large one-vat workshop needed about 200 wall "faces" (*mian*) of one by two meters, which had to be repaired every year. If we include the cost of drying walls, total investment in a workshop amounts to the net income of a workshop over three to four years. On the other hand, equipment could be acquired step by step; workshops could start with a rented vat and a few drying walls and expand over time.

The first prerequisite for workshop expansion was a favorable demographic balance in the household. As the Russian economist Aleksandr Chayanov demonstrated, economic differentiation in rural economies depends to a large extent on fluctuations in the worker-consumer ratio in the household. Families with young children have to work hard just to feed themselves. As the children grow up and start working, the household farm or workshop expands, until the family splits up and the cycle begins again. In the paper industry, this process is complicated by the inflexible labor demands of the vat. As explained in Chapter 1, a one-vat workshop needed six or seven workers to achieve full productivity; smaller workshops could not ensure that the core workers (pulper, vatman, and brusher) remained at their posts for ten to twelve hours a day. Since productivity in small workshops was low, they found it difficult to hire the workers necessary to reach and cross that threshold. Workshops tended to rise on the crest of a demographic tide: with two grown sons, or a son and a daughter-in-law, workshops could establish a division of labor in which each adult concentrated on a single task. The higher productivity that followed allowed the workshop to hire additional labor. This was done not simply to replace household with hired labor but to adjust the internal division of labor in ways that helped move the workshop toward higher market segments. Hiring a vatman, for example, freed a son to learn how to prepare pulp, a crucial skill for any future workshop owner. This in turn freed the father of the family to devote most of his time to quality control and marketing. With better product quality and with personal contacts to wealthy paper traders, he could hope to join the ranks of big producers.

Informants who owned paper workshops before 1949 agreed that "management" (*guanli*), rather than technical skill or capital, was the key to success. This was partly because of the fragmented nature of the paper market, discussed in more detail below. To put it simply, small workshops produced "small paper" (*xiaozhi*), paper in small sizes and of comparatively poor quality that was used for wrapping or burned in ancestral sacrifices. Small paper was a generic product, bought by itinerant buyers (*zhi fanzi*) and sold via intermediaries to distant markets. Profit rates in this market were low and depended on cost competitiveness rather than quality. By contrast, large-format printing and writing paper (*dazhi*) was a brand-name product, stamped by the wholesale paper shops or (more rarely) the producer, who vouched for its quality. Since it was in practice impossible for buyers to check the quality of every sheet of paper, sales in this market sector were based on trust between buyer and seller. Producers who enjoyed the "trust" or "credit" (*xinyong*) of buyers could ask as much as 30 percent more for the same batch of paper than sellers not known to or not trusted by the buyer.[3] Often, it was difficult for small producers to find buyers, since large paper shops, which paid the highest prices, were reluctant to deal with workshops that could not guarantee a steady output of paper of consistently high quality. In order to succeed, then, a workshop owner needed to cultivate connections with buyers, which meant spending a great deal of time in the county seat or market towns.

The need to cultivate stable relationships with buyers and the size of the initial investment presented formidable obstacles for young households wanting to establish themselves. Former small producers and wage workers recall a highly polarized ownership structure in Shiyan, with little mobility between the dominant "big households" and the rest. One telling detail is that by the 1940s, sons of big workshop owners no longer learned to mold because they assumed that others would do the heavy work in their workshops.[4] Mobility was not, however, as uncommon as this suggests: Shi Ziqing, the most successful papermaker of Republican-period Jiajiang, started with a small and not very profitable workshop and ended with a permanent workforce of a hundred workers and a fixed capital of 20,000 silver *yuan*; Shi Haibo managed to expand the one-vat work-shop he inherited to five vats. On a smaller scale, Shi Yuqing saved enough money in five years of wage work to start a workshop, to which he soon added a steamer of his own—all this before he turned 25. Such success stories, together with the edifying spectacle of rich families ruined by laziness, gambling, or opium addiction, convinced people in the paper districts that wealth and poverty rarely lasted longer than a generation.[5]

During the expansion of the paper industry in the 1930s and early 1940s, skilled labor was in short supply. Young vatmen commanded premium wages in Hedong; older men no longer strong enough to mold large-format paper often hired out to small-paper workshops in Hexi. Contrary to the assumption that Chinese custom kept women locked up in their homes,[6] women in Jiajiang could and did hire out for wages. Unmarried women did not hire out, although the informal swapping of drying walls between households meant that women often found themselves working in the courtyards and houses of their neighbors. Married women, however, worked routinely for relatives or neighbors or even hired out to distant places. Most wage brushers in the Hexi industry were from Hedong, and although many of them worked in teams with their husbands, others found work on their own.[7] Unskilled men and women could find work as porters; young men looking for light work could guard bamboo groves against thieves. As discussed in the previous chapter, wages in the industry were relatively high: a vatman earned three to six *jin* of rice a day, enough to feed three to six persons; a brusher about half that amount.[8] Even unskilled workers could earn enough to feed "two and a half persons, although not three." By the standards of Republican rural China, these were decent wages, yet former wage workers also remember poverty and insecurity. Although workers had enough to eat even in times when business was slow, they owned little property: some land perhaps (unirrigated land was cheap), but often no more than one or two pieces or furniture, one blanket for the entire family, and one or two sets of patched clothes per person. What was worse, perhaps, was the fear that illness might put them out of work. Shi Rongqing, who started molding when he was twelve and later worked as a vatman for Shi Ziqing, remembered that often his hands trembled so badly that he could no longer mold, and that when he did not mold, he and his parents had nothing to eat.[9]

Relations between workers and employers were usually amiable, for the simple reason that if they were not, one side would terminate the contract. Big workshops like that of Shi Ziqing offered the best conditions: high wages, paid when needed (often in advance); good food; steady and predictable employment. Food was considered an important part of the wage, and conflicts between workers and employers often centered on the quantity and quality of dishes. In the 1930s and 1940s, meat was customarily served on the second and sixteenth day of each lunar month, and poor fare—for example, pork belly that had been rendered down before serving—or delayed meat days could lead to loud complaints or indeed to a change of employers. Another source of conflict was the final wage calculation at the end of the year. Most workers kept open accounts with their employers, drawing cash or grain advances as needed. Accounts were settled just before New Year's or when a worker changed employers, and workshop owners sometimes docked wages for damaged tools or shoddy work. Yet Shi Haibo, owner of the second largest workshop in Shiyan, explained that it was in the interest of employers to keep their workers happy: "If you didn't get along with your workers, you couldn't make any money." Strict supervision was costly and ineffi-cient; in order to make a profit, employers had to motivate their workers to maintain certain quality standards. If they antagonized their workers, they would waste raw materials or produce shoddy work in ways that could not be detected until it was too late.

Customary obligations also governed the relation between "big households" and their neighbors. Shi Haibo related that he allowed his poorer neighbors to use his steamer and other large equipment. When he steamed his own *liaozi*, he often left one-third of the steamer empty and allowed his neighbors to fill up the remaining space, thus saving them expenses for quicklime, soda, and fuel. He made loans to poor neighbors, knowing that they could not repay them, and employed needy people even though they were slow workers. Like other large producers, he contributed to communal expenses: when a group of poor neighbors decided to build a steamer, they provided the labor and Haibo paid for the raw materials. More important, he, like other owners of big workshops, sponsored the pa-per sales of his neighbors. Since large paper traders were unwilling to buy from small producers, the latter often entrusted their goods to "big households," who sold them under their own name and passed on the payment to the original producer. Since the producers were not present during the sale, they could not verify if they indeed re-ceived the full amount, but people from both sides claimed that such deals were sanctified by custom and that cheating was rare.[10] Help-ing one's neighbors, Haibo explained, was not a matter of individual charity; failure to comply could lead to the kind of low-key yet pain-ful penalties that James Scott described as the "weapons of the weak"—slander, gossip, social ostracism, and so on.

A "Big Household": The Workshop of Shi Ziqing

Although not the largest paper producer in Republican-period Jiajiang, Shi Ziqing (1894–1938) was probably the most prominent, with a biographical entry in the 1989 county gazetteer and other local publications dedicated to his memory.[11] In Shiyan, Ziqing is fondly remembered as a model employer and a patron of the Shi community, but he owes his fame largely to his association with famous men of culture (*wenren*) such as the painter Zhang Daqian. In 1913, Ziqing inherited a small and unprofitable workshop from his father. After learning pulping techniques from his father-in-law, he specialized in making large-format *lianshi* paper of the highest quality. At the height of his success, in the mid-1930s, Ziqing's workshop consisted of twelve soaking basins, eight paper vats, and five steamers, with a total investment of around 20,000 silver *yuan*. He employed "ten tables" of skilled workers (i.e., 80 workers); during harvest and steaming time, employment rose to around 200.

The accounts of Ziqing's life stress his attention to quality and his willingness to introduce new technologies, but his success was due largely to his business acumen and personal flair. Like other producers of "big paper," Ziqing worked hard to establish a brand name for his paper and to prevent counterfeiting. He not only stamped each ream of paper on the edge and outer wrapping but also watermarked his paper, overcoming considerable technical odds.[12] His first success came at the 1920 Chengdu trade exhibition, when he was awarded a horizontal plaque (*hengbian*), inscribed in the calligraphy of the provincial governor with the phrase *bao wo fuyuan*, "protect the sources of our wealth." A few years later, he won another board inscribed *huanhui liquan*, "restore [our] rights and interests."[13] Although only semi-literate, Ziqing actively sought to associate his paper with themes of patriotism, Chineseness, and national culture. More than other producers, he courted painters, calligraphers, and literati, inviting visitors to his house in the county seat or to his large courtyard house and workshop, spectacularly located on a cliff overlooking the Shiyan valley. The outbreak of the Sino-Japanese war further boosted the fortunes of his workshop, although Ziqing himself died before the beginning of the wartime boom. As "patriotic artists" such as Qi Baishi, Xu Beihong, and Zhang Daqian relocated to southwestern China, they found themselves cut off from supplies of their favorite *xuan* paper from Anhui. They turned to Jiajiang paper as a second-best alternative but remained dissatisfied with it until Zhang Daqian teamed up with Ziqing's son, Guoliang. For two weeks, Zhang moved in with Guoliang, spending the days in the workshop and the nights (according to local gossip) drinking and smoking opium, until they came up with the right mixture of fibers.

His biographers describe Ziqing as a practicing Confucian who declared that "his success would not be complete until all neighbors had work to do and rice to eat," who advanced wages to his workers, tolerated bamboo pilfering (as long as it remained within customary limits), and gave generous grain and cash loans to workers and neighbors. Among the Shis, Ziqing is remembered as a friendly, chubby man, so frugal that he picked up bamboo twigs dropped by the carriers, but generous to children (several informants had received cash gifts from him) and strangers. He is best remembered, however, as a champion of the Shis' economic interests in their long-standing rivalry with the Macun Mas. In 1936, Ziqing and other wealthy Shis built a paved market street with two rows of houses in Jiadangqiao, in the heart of the area inhabited by the Shis. Market streets were commercial investments, but Ziqing's main motivation was to "facilitate the life and work of papermakers" in Shiyan and reduce their dependence on the market town of Macun. Jiadangqiao soon developed into a substantial settlement with around forty houses, many shops, and a teahouse, although the periodic market collapsed in 1947 and never revived.

Like most big producers, Ziqing treated his workers well. Wages were higher than elsewhere, and meat was served almost every week; vatmen received meat and liquor every other day. In contrast to Guoliang, Ziqing did not smoke opium and helped workers to give up the habit by paying their wages during detoxification. He often gave wage advances and cash loans, and rarely pressed for repayment. However, even this model employer did not escape conflict. Although Ziqing was popular with his workers, his wife, who was in charge of recordkeeping and financial management, was called the "heaven-blocking king" (*dangtian wang*) because of her strictness.

On one occasion, workers carried freshly harvested bamboo to the workshop, and Ziqing's wife recorded the number of delivered loads. The workers soon noticed that she never looked up but simply made a mark each time she heard a bundle fall. They hid behind a corner, smoked their pipes, and from time to time lifted a bundle and dropped it again with a loud thump. The next day, the Heaven-Blocking King discovered that she had been cheated and ordered the kitchen to mix the rice for lunch with broad beans. The workers grumbled and spat the bean skins on the floor. Now thoroughly incensed, the Heaven-Blocking King ordered her helpers to sweep up the bean skins and bring them back to the kitchen. At dinner, the workers were served deep-fried bean skins. Perplexed, they said: "Ah, deep-fried bean skins! Now that is a good meal!" The man who told me the story was himself unsure who had won this contest of wills.

Class and Landownership

For papermakers, landownership was a result rather than a precondition of economic success. Successful papermakers bought bamboo land to ensure stable supplies of raw materials, and large workshops such as Shi Ziqing's bought paddy fields in the plains and fed their workers with the 30–40 percent of the harvest that tenants paid in rent. In Shiyan, where landownership was more concentrated than elsewhere in the paper districts, the top 10 percent of landowners owned 38 percent of arable land. Even in this group, however, the average holding per household was only 6.5 *mu*.[14] Ownership of bamboo land was more concentrated: the top 10 percent of owners held 68 percent, with an average holding of 28 *mu*. Papermakers liked to own bamboo land since this reduced production costs, but it was not essential; almost all workshops relied to some extent on bamboo purchased in the market. Tenancy was almost unknown in the Hedong paper districts: arable land (almost all of it on steep hill slopes) was too poor to produce much surplus, and bamboo land was rarely rented out. In Hexi, land was easily available; at the same time, tenancy was more common and came with almost feudal burdens. Whereas poor households in Hedong could always earn a living in the paper industry, the poor in Hexi had no choice but to rent land from landlords, paying up to half of the harvest in rent. Contracts lasted from one to ten years and could be terminated at the whim of the landlord. Tenants were expected to give gifts to landlords during major festivals and on occasion to provide unpaid corvée labor. Rich landlords in remote Hexi were also more likely to use political power to enrich themselves. Xie Changfu remembered that Liu Zhihua, head of Huatou township in the 1940s, contracted bamboo land from all the temples in Huatou and ordered his tenants to harvest his bamboo, giving them only food for their work. Since Liu could order people drafted into the army, nobody dared complain.[15]

Markets

In 1685, when the Jiajiang paper industry first appears in the historical record, Jiajiang paper was sold mainly in Chengdu and along the upper reaches of the Yangzi, in an area stretching from Leshan, Luzhou, and Yibin to Chongqing and Wanxian.[16] These two regions remained the most important markets for Jiajiang paper until late into the Republican period when new markets were opened in the southwest—in Yunnan, Guizhou, and the short-lived province of Xikang (which included parts of western Sichuan and eastern Tibet)—and in the northwest, in Gansu, Shaanxi, and Shanxi. In Republican times, the major markets for Jiajiang paper were Chengdu (40 percent of the output), Chongqing (18 percent), Yibin (10 percent), Luzhou (9 percent), Kunming (8 percent), followed by Xi'an, Lanzhou, and Taiyuan.[17] In the 1930s and 1940s, over a hundred Jiajiang paper merchants were active in Chengdu, and about a dozen each in Chongqing, Yibin, Luzhou, and Kunming. These places were not necessarily the final markets: Chengdu, apart from being the largest market for Jiajiang paper, was also a distribution center for the north Sichuan hinterland; Chongqing served the same function for eastern Sichuan, as did Luzhou and Yibin for southern Sichuan, Yunnan, and Guizhou. As elsewhere in China, paper markets were fragmented: for example, *duifang* (a medium-size, medium-quality writing paper) sold in Chengdu but not in Chongqing, where the more expensive *lianshi* was preferred; high-quality *gongchuan* sold in Kunming but not in Chongqing or Chengdu. Some products sold only in one or two counties: five-colored *pingsong* was marketed in Qiongzhou; "foreign small blue" (*yangxiaoqing*) in Ziyang and Neijiang (see Appendix C for a list of paper types and their markets).[18]

Although most paper left Jiajiang in unprocessed form, some was dyed or printed locally. In the Republican period, Jiajiang city was home to 40 to 50 dyeshops (*ranzhifang*), also known as "red paper workshops" (*hongzhi zuofang*), even though they dyed paper in all colors. Dyeshops also sprinkled paper with specks of gold, coated it with tinfoil, or marbled it in a "tiger skin" pattern. By the late Republican period, the number of dyeshops had increased to 270, processing some 2,000 tons of paper every year. "Paste shops" (*tiezadian*) catered to the immense demand for paper money and funeral goods. In these shops, paper was folded into imitation silver ingots or strings of cash, stamped with prayers or petitions to the gods, tailored into funeral clothes and shoes, or pasted on bamboo frames to produce houses, sedan chairs, farm animals, and other goods that were burned to accompany the deceased into the afterworld. The paste shops also made stationery (envelopes, writing pads, account books, invitation cards), kites, lanterns, and paper toys. About twenty small print shops produced New Year's pictures (*nianhua*), door gods (*menshen*), and matching couplets (*duilian*) to decorate walls and door frames at New Year's. Commercial publishers printed the Confucian classics, history books, textbooks such as the *Three Character Classic* and the *Classic of Filial Piety*, almanacs, opera libretti (the Mingxintang publishing house in Jiajiang alone produced more than 50 of these), collections of folk songs (*shan'ge*), fairy tales, and popular novels. By 1949, Jiajiang had ten modern machine printshops that produced advertisements for companies in Chengdu, Yibin, and Leshan. Finally, the paper industry stimulated the growth of other, related trades. About ten workshops in Jiajiang city produced brushes, sixteen made ink; in addition, there were, among other trades, screen makers who produced molding screens, blacksmiths who made razor-sharp paper knives, masons who made molding vats and basins, packaging firms, shipping agencies, and importers of bamboo and soda.[19]

Before the building of motor roads in the 1920s and 1930s, paper was transported on boats when possible and on human backs or carts when not. In the hills, porters (*lifu*) used bamboo carrying poles to transport heavy loads on narrow paths. Most of the flagstone roads in the plains were wide enough only for wheelbarrows; after the motor roads were built, wheelbarrows were replaced with large carts (*banche*) that could carry up to two tons and were pulled by teams of up to ten coolies. However, most long-distance transport was on water. The Qingyi river at Jiajiang was navigable by small craft and bamboo rafts that were floated down from Ya'an on the Tibetan border and could carry up to eighteen tons.[20] In Leshan, goods destined for Chengdu were transshipped and towed up the Min River; goods for all other destinations were shipped down the Min to Yibin, and from there down the Yangzi.[21] Whether by land or water, transport was painfully slow. Zhai Shiyuan, a Jiajiang native who sold paper in Kunming, described the hardships of life on the road in the early 1930s:

The paper was shipped to Yibin and from there to Hengjiang, where it was unloaded. We then hired porters up to Laoyatan, and from there muleteers up to Shaotong and Kunming. . . . There were twenty-odd stations on the way; one station was one day's journey. If we missed it, we had to sleep out in the wild. I marched with the porters and muleteers, taking only brief breaks on the way and then pressing on. After about 50 days we arrived in Kunming, tired and with swollen feet. During daytime, we only ate cold rice, with salt instead of side dishes. . . . At that time, traders who sold Jiajiang paper in Kunming went back and forth twice a year.[22]

Transportation improved rapidly in the Republican period. The first steamboat crossed the Three Gorges in 1898; regular services between Yichang in Hubei province and Chongqing were established in 1909. In 1914, steamers reached Leshan, and by the 1940s, most major cities in Sichuan were linked to Chongqing by steamer. However, most steamers carried only passengers; commercial goods relied largely on junks, sometimes towed by steamers.[23] Overland transportation also improved markedly, as competing warlord governments built military roads.[24] After the reintegration of Sichuan with the rest of China in 1936, the various warlord-built road networks were linked and joined to those in neighboring provinces.[25] In 1928, a daily bus service linked Jiajiang to Chengdu and Leshan, and in the 1940s, trucks plied the Jiajiang–Chengdu route to bring paper to Chengdu.[26] Paper traders responded quickly to improvements in transportation. When the Chengdu–Xi'an motor road was completed in 1937, Xi'an emerged as a major market; the next year, when the Chengdu–Kunming road was opened, Kunming became the third-largest market.[27]

The market structure for Jiajiang paper was in many ways a mirror image of the production structure. Most trading firms, like most workshops, were household-based and undercapitalized; traders, like producers, were enmeshed in networks of mutual obligation. In contrast to the situation in many other industries, the marketing of Jiajiang paper lay in the hands of Jiajiang natives rather than visiting merchants. Most trading firms consisted of a purchasing station (usually a small shop) in Jiajiang and an outlet (a shop or warehouse) in the final urban market. In 1937, the chairman of the Jiajiang Paper Trade Association estimated that there were several thousand Jiajiang traders in the province. This seems too high, but a number of around a thousand appears possible.[28] In Chengdu alone, there were more than a hundred Jiajiang paper traders, most of whom operated out of warehouses (*duizhan*) near the Min River wharves. These warehouses were subdivided into small, boxlike compartments, in which paper traders cooked, ate, and slept, surrounded by piles of paper.[29] Retailing was left to Chengdu natives, who bought paper from the Jiajiang traders. A similar division of labor could be found in other cities: in Kunming, only one of ten Jiajiang traders had a retail store; all others rented warehouse space; in Yibin, five of twelve Jiajiang traders opened retail stores.[30] According to Ren Zhijun, who sold paper in Chengdu and Kunming in the 1940s, Jiajiang traders obeyed a strict business code in their dealings with one another and with local buyers. One's credit or reputation (*xinyong*) was one's "second life"; anyone who lost it "could not survive as a trader." Written contracts were rare, since most transactions were between people who knew and trusted one another. Only when dealing with strangers would traders ask for a deposit or go through middlemen.[31]

Purchasing operations in Jiajiang were as small-scale and dispersed as urban retail markets. A 1943 study lists several ways in which paper was collected from producers: itinerant purchasing agents (*zhi fanzi*) who scoured the paper districts, agents who introduced sellers to buyers for a small commission, and three different types of resident paper merchants: "paper shops" (*zhipu*), "local transport traders" (*bendi fanyunshang*), and "outside purchasing agents" (*wailai caigoushang*). Paper shops, of which there were about a hundred, collected paper to sell it to other traders. Local transport traders, of whom there were 400–500 in the county seat and in the market town of Mucheng, were the purchasing stations for urban shops and warehouses. Purchasing agents were representatives of large merchant firms and publishing houses in Chengdu and Chongqing. There were only five of them, and they bought large quantities of expensive print and writing paper from "big households" such as Shi Ziqing. Another option, not discussed in the 1943 study, was to visit one of the six paper markets in the county—at the North Gate and West Gate of the county seat, and in the market towns of Mucheng, Huatou, Macun, or Shuangfu—and buy directly from small producers. Finally, there were dyeshops, book publishers, and producers of funeral goods, all of which had their own supply sources.[32]

Generally speaking, large merchants dealt with large producers with whom they built up long-term, stable, and mutually beneficial relationships. Small traders, by contrast, dealt with small papermakers, and transactions tended to be short and impersonal, with none of the mutual trust and gentlemanly leisure that characterized the high end of the trade. A typical example of a relatively large merchant firm is that of Xu Shiqing, whose father and grandfather owned a paper shop in Yibin and a purchasing station in Jiajiang. Xu, who later became a cadre in the county Light Industry administration, remembers that his father advanced cash to producers when they needed it, charged only moderate interest on loans, and offered loans to producers who could not deliver on time. Sometimes this meant throwing good money after bad, but the Xus had learned from experience that paper workshops rarely went bankrupt; sooner or later, they recovered from their problems and repaid their debts. In any case, the social and financial costs of debt enforcement were too high. For firms like Xu Shiqing's, reputation was the most important asset: respectable merchants did not chase after small profits and accepted occasional losses with good grace.

Small traders, by contrast, dealt with small workshops that were more likely to default. Cash-short producers sometimes promised their paper to several buyers or accepted advance cash or raw materials from one trader only to sell their paper to another. Small traders therefore insisted on immediate payment; if they gave loans at all, they charged high interest rates. In Hexi, only a few large workshops had permanent credit relations with established merchants. Sideline producers either sold their goods to itinerant buyers who went from door to door or (the least preferable option) took their paper to the markets of Jiajiang, Huatou, and Mucheng. Papermakers who had walked many hours to the market had to sell their paper before noon or spend the night in town, since no one dared to travel after nightfall. Buyers knew this and waited until the afternoon, when prices came tumbling down. It was in order to avoid such transactions that small papermakers entrusted their wares to their more successful neighbors, asking them to sell on their behalf.

Credit

A full production cycle, from harvesting the bamboo to carrying the finished paper to the market, took about three months. In order to bridge the interval between capital outlay and returns, almost all paper workshops borrowed money from paper traders. Large workshops that maintained a steady output were less dependent on such loans, but even they borrowed money to increase their working capital. Sources from the Republican and early PRC period describe credit arrangements in the paper industry—locally known as *yuhuo* (advance sale) or *xia cao* (reserving a vat)—as extremely exploitative, and indeed, many paper workshops remained indebted to the same merchant firm for years, sometimes for generations. Yet this is not how contemporary informants remember the credit system. In their view, credit—almost any form of credit—meant security: a creditor was unlikely to cut off credit to a debtor, and repayment could always be renegotiated or delayed. To be in debt meant to be in business; the really unfortunate were those who were deemed such poor credit risks that they could not obtain loans and were thus forced to sell their goods *xianhuo*—in impersonal, cash-for-paper exchanges in the open market.

Yuhuo relationships were typically initiated by papermakers. An aspiring seller would bring paper samples to the market and show them to potential buyers. If a would-be seller had no previous contacts, he employed a middleman. Buyer and seller would then negotiate price and time of delivery. This was done orally or in a brief and informal written contract. If the trader advanced cash against delivery, the recipient signed and fingerprinted a receipt. After one or two seasons, such one-off loans evolved into open accounts: workshops would ask for as much cash or raw material advances as they needed and pay their debt back in kind as soon as the paper was finished. Accounts were settled at the end of the year; if this was impossible, the balance was carried over into the next year. For most of the year, the balance was negative for the producer, but the reverse could also be the case, for example, when paper traders took paper on commission and paid for it only after completing their sales. In long-established relationships, both sides would tolerate overdrafts, without charging interest.[33]

Interest was charged mainly in short-term transactions between traders and producers who did not know each other well. In these cases, the trader would advance cash early in the season and be paid back in paper later in the season. Rather than charging monthly interest, he would deduct a certain sum, customarily 10–20 percent, from the sales price. If, for example, buyer and seller had agreed on a price of 100 *yuan* for 10,000 sheets of paper, the buyer would pay 80 *yuan* and treat 20 *yuan* as interest on his loan. For the lender, this meant that he paid 20 *yuan* in interest, or 25 percent, on a loan of 80 *yuan*. Monthly interest rates thus depended on the duration of the loan: if the producer delivered after three months (the most common duration of such loans), the monthly rate was 8.3 percent; if after four months, it was 6.25 percent. This may appear high, but it was much lower than the 30 percent a month often charged for commercial loans. Instead of a cash advance, a papermaker might ask for soda, bleach, or other raw materials from a paper shop (many paper shops also sold raw materials) or from a specialized supplier.

The mechanism here was similar: if the trader knew and trusted his counterpart and expected swift repayment, no interest was charged; if repayment was expected to be slow or uncertain, traders would add a surcharge of 5–20 percent to the price of their goods.

Interest rates are commonly understood to compensate for inflation losses, risk of default, and lost alternative investment opportunities while the money is tied up in the loan. Since the loans in the papermaking industry were repaid in kind, not in cash, inflation did not matter: a trader who paid a certain sum for 10,000 sheets received exactly the 10,000 sheets he had paid for, regardless of changes in the price of the paper. The risk of default was a more important consideration. Elderly paper traders and producers I interviewed agreed that a debtor who was determined not to pay could not be forced to do so. The state's arm did not reach into the villages, and other authorities—lineage councils or the ubiquitous "robed brothers" (*paoge*)—were reluctant to get involved in disputes between local producers and outside traders. Outright default was relatively rare, but papermakers in dire straits sometimes accepted cash payments from several buyers and fulfilled their obligations to only one of them or simply delayed the delivery of their paper, which effectively reduced the cost of their loan. A papermaker who defaulted or delayed payment risked losing his access to credit, which was serious enough, but he risked little else.[34] Lost returns on alternative investments were considerable: short-term loans outside the paper trade earned a flat rate (known as a "knocking interest," *qiaoqiao li*) of 33 percent per month or a compound rate ("rolling interest," *gungun li*) of 30 percent per month. In both cases, a lender could double his investment in three months.

A fourth factor that needs to be taken into consideration was the seasonal fluctuation in the prices of raw materials and paper. Sources from the 1930s to 1950s likened *yuhuo* to buying grain "on the stalk" (*mai qingmiao*)—a usurious practice in which grain merchants bought as yet unharvested grain in early spring, when cash-starved farmers were willing to accept low prices. Paper merchants, the critics argued, took advantage of capital shortages in the summer season, forcing producers to sell cheap and buy dear. However, the analogy with the grain market does not hold. Paper prices did fluctuate but not seasonally. A workshop that sold its paper in August at August prices and delivered in September might benefit or lose, depending on how the price developed.[35] The situation was somewhat different for raw material prices, which fluctuated in a predictable seasonal pattern. Bleach, bamboo, and soda were 25–50 percent higher in the busy season than in slack periods. Yet prices peaked in different months for different products: soda was most expensive in August, during the busy steaming season, whereas bleach was most expensive from September to December, when molding was most intense.[36] Loans allowed papermakers to spread out their expenses and to buy when prices were relatively low. All my informants agreed that the worst situation—far worse than long-term debt—was to be denied credit, to have to buy and sell for ready cash (*xianhuo*) in an impersonal market setting. *Xianhuo* transactions were uniformly described as exploitative and risky, the resort of those who had no other choice.

A 1943 report on the Jiajiang paper industry estimated that 70 percent of Jiajiang papermakers practiced *yuhuo*; only very small workshops sold all their paper for cash.[37] For the most part, working capital—outlays for bamboo, bleach, soda, and food grain—came from commercial loans; most fixed capital came from the papermakers themselves. A rough estimate of the credit volume in the industry can be derived from studies of industry in the 1940s. Zhong Chongmin's 1943 study estimated an annual capital demand of 12,690 *yuan* per vat, not including the costs of household labor. Of this, roughly 25 percent was for bamboo, 15 percent for soda and quicklime, 20 percent for bleach, 30 percent for hired labor, and 10 percent for depreciation and minor expenses. The total yearly demand for the 2,937 vats in Jiajiang was thus around 37 million *yuan*.[38] One year and another inflation spurt later, Hua Younian reported that remittances from paper shops to their purchasing stations in the half year between July and December 1943 amounted to 80 million *yuan*. Leaving aside remittances in the (less busy) first two quarters of the year and loans that reached paper shops through other channels, and assuming that 80 percent of this sum reached the workshops (the rest being spent on transportation and operating expenses of the paper shops), a minimum estimate of the credit volume would be 64 million *yuan*. The average yearly credit volume for each vat would then be 21,791 *yuan*, equivalent to 763 kg of rice. We can thus tentatively assume that on average, a one-vat workshop received loans equivalent to three-quarters of a metric ton of rice each year.[39]

Market and Community in the Paper Districts

Since publication of G. William Skinner's classic studies, marketing has been seen as central to the understanding of Chinese social structure.[40] These studies are particularly pertinent to the Jiajiang case, since Skinner's fieldwork took place in the Chengdu basin, not far from Jiajiang, and his model reflects the peculiarities of marketing in Sichuan. Skinner's argument—similar to Freedman's in its all-encompassing scope—is based on the convincing premise that marketing creates a matrix of routine economic links, which can then be mobilized for a wide range of social functions. Grain and other agrarian products move upward to the cities, and retail goods move downward to the villages, through a multitiered hierarchy of markets. At the basis of the structure, standard markets serve as retailing centers and collection points for the surplus of the surrounding countryside; at its apex, each market hierarchy has a central metropolis where all marketing ties come together. These marketing hierarchies are sharply distinct from one another; Skinner defined nine macroregions, each of which includes parts of several provinces. These market hierarchies form "the 'natural' structure of Chinese society—a world of marketing and trading systems, informal politics, and nested subcultures dominated by officials-out-of-office, nonofficial gentry, and important merchants."[41] Although Skinner's focus is on retail marketing, he argues that the same hierarchies "are optimal . . . also for the collection and export of local products, for the import and distribution of exogenous products, and for the wholesaling, transport, and credit functions essential for these activities."[42]

Skinner's central assumption—that social networks and organizations rest on an underlying matrix of market ties—is borne out by the evidence on Jiajiang guilds and other networks, discussed in the next chapter. Yet as Carolyn Cartier has pointed out, Skinner's environmentally deterministic model leaves little room for human agency. Like Freedman's lineage paradigm, it sees rural people as following a set of fixed principles in an unconscious and predetermined fashion. At the most basic level of the model, there is no freedom of choice: people do all their marketing in "their" standard market; even if they live at equal distance to two market towns, they use only one of them. Freedom of choice increases for actors at a higher level of the hierarchy: businesses in a standard market town, for example, have links to two or three higher-order markets. However, at all levels of the model, human activity remains constrained by an economic rationality that operates quite independently of individual will. This assumption is contradicted by what I found in Jiajiang. According to my informants, people exercised free will in their choice of markets and often acted in ways that appear irrational in strictly economic terms. People might dislike certain markets for personal or political reasons and often walked long distances to avoid them; or, if they could not avoid them, they kept social intercourse to a minimum. Even for the sale and purchase of bulk goods such as bamboo, soda, and food grain, people did not necessarily go to the nearest market, and transport costs, though undoubtedly high, did not have the constraining impact that Skinner predicts. This is even more true if we look at the marketing of paper, a high-value, lightweight good that could bear high transport costs.

As Skinner has shown, market schedules tell us how markets are coordinated with one another—subordinate standard markets avoid a schedule that clashes with the higher-order intermediate market on which they depend. Jiajiang county had fourteen markets in thirteen different towns (the county seat had two separate markets). Jiajiang city and the important market town of Mucheng on the Qingyi River stand out as intermediate markets. Jiajiang city had a *dan* schedule: its markets convened on every odd day of the lunar month; predictably, market towns near the county seat (Ganjiang, Yongxing, Chafang, Tumen, Xinxin, and Macun) adopted a *shuang* (even-day) schedule. Three markets near the county border were synchronized with higher-order markets in neighboring counties. The market of Mucheng, to the north of the county seat, met on days 2, 4, 6, and 9 of the ten-day market cycle. All markets in the paper districts with the exception of Macun (Nan'an, Yingjiang, Zhongxing, Xiema, and

Huatou) were synchronized with Mucheng, reflecting its importance as the main transshipment point for paper and as the main port for grain, bamboo, and other inputs.

This may sound as if markets were relatively static, but the reverse was true. Chinese markets were businesses, owned and operated by corporate investors (lineages or groups of merchants) who rented out shop space and collected fees from stalls. I have mentioned the Huatou market, founded in 1646 by the Xiaos, Luos, and Guos, who built a market street and a wharf on the Yachuan creek. Two hundred years later, the Xies and Tangs added a second street. These were long-term investments: the Xiaos owned the first market and the wharf, and levied fees on them; the Luos levied slaughter fees; and the Guos collected a surcharge on grain.[43] In a similar fashion, Shi Ziqing, Shi Longting, and other Shi investors opened a market at Jiadangqiao in 1936. Markets for special products could be added to existing markets: Macun, for example, added paper, wax, and silk markets to its periodic grain and vegetable market in the 1900s, but closed them after a devastating raid by the bandit Monk Zhang and his gang.[44] Market-making was not always motivated by purely economic aims. Shi Ziqing and his associates clearly hoped to earn a return on their investment in the Jiadangqiao market, but they also wanted to reduce the dependence of the Shis on Macun and the Mas. Similarly, the Shuangfu market on the border between Emei and Jiajiang was established by Emei traders who hoped to divert business from Jiajiang to their hometown. This was immediately countered by Jiajiang paper traders, who opened a new market in Dugong, on the Jiajiang side of the border, and underbid their Emei rivals until they closed down. Once the Dugong market had fulfilled its purpose of pushing out the Emei competition, it, too, was closed.[45]

Skinner argued that "insofar as the Chinese peasant can be said to live in a self-contained world, that world is not the village but the standard market community."[46] This has been disputed for other parts of China. Philip Huang, for example, found that non-elite villagers in north China rarely interacted with people from other villages during their visits to market towns.[47] The situation in Jiajiang seems to have resembled Huang's north China villages; none of my informants reported the "nodding acquaintance with almost

every adult in all parts of the marketing system" that Skinner observed in his fieldwork site.[48] People from Shiyan went to Macun to
buy vegetables, condiments, and household goods, but few people
spent much time in the town, and most other business was conducted in the county seat or in Mucheng. One reason for this was
the high price of grain in the Macun market: like the rest of the paper districts, Macun was short of grain, and the local grain market catered only to the needs of townspeople. Papermakers from Shiyan
used to walk up to 20 km to grain markets in the plains and return
with loads of roasted rice (*huomi*) on their backs.[49] One load of 45 kg
would last a family of five for about ten days. In Hexi, too, papermakers seldom frequented their local markets. Most grew enough
grain for about half the year; during the rest of the year, they bought
rice in the plains, where it was cheaper. Large workshops often owned
paddy fields in the plains and received rent in kind; those that did not
ordered their grain from wholesalers (*mi fanzi*) who delivered it to the
doorstep. Since wage workers ate in the workshops and were often
paid in grain, they did not need to go to the market.[50]

In addition to grain, workshops needed large quantities of bamboo, fuel, and soda. Most of these goods were ordered in Jiajiang or
Mucheng, not in the standard markets, and delivered to the workshop door. Although Macun, Huatou, and Shuangfu had small paper
markets, most paper was delivered to traders in Jiajiang and Mucheng. Big workshop owners went to these towns about once a
month, accompanied by their strongest workers, each carrying 60 kg
of paper on a carrying pole. After selling their goods, workshop owners sent the workers back, loaded with salt, tobacco, cloth, and chlorine bleach, while they themselves stayed behind to relax with
friends in teahouses or restaurants. Small papermakers, by contrast,
carried their goods to town and returned quickly, so as not to be
tempted into making unnecessary purchases.[51] The relative distance
of papermakers from their market towns is also reflected in marriage
markets. The Shis frequently intermarried with their neighbors, but
they also recruited brides from other townships and even from outside the county. Ideally, brides were recruited not through professional go-betweens, who were believed to be greedy and unreliable,
but through female relatives. Mothers looking for a daughter-in-law

often turned toward relatives in their own natal village. Elderly people in Shiyan estimate that about one-third of all brides came from outside the township and perhaps 5 percent from outside the county.[52]

People in Shiyan had an ambivalent relationship with their market town, Macun, which was dominated by the Mas, with whom the Shis were frequently at odds. Until the 1930s, papermakers from all parts of the township attended the yearly Cai Lun festival in Macun, held in honor of the trade god of papermaking. Yet Macun was not a papermaking town, and institutional life in Macun focused on agriculture rather than on the paper industry. The main annual festivals centered on the Buddhist Golden Dragon temple and on the temple of Chuanzhu (Lord of the River), the protector god of Sichuan, associated with irrigated agriculture, state power, and territorial control.[53] In the Republican period, Macun developed a vibrant associational life, with three *paoge* lodges and a proliferation of voluntary associations, mostly linked to the dominant Jiang faction.[54] Most Shis, by contrast, sympathized with the more conservative Wang faction, whose leaders were long-established paper traders. A former wage worker from Shiyan remembered that he used to go to Macun—less than an hour's walk away—about twice a week, to buy food and "watch the excitement" (*kan renao*) on noisy market days. But when I asked whether he had friends in Macun or participated in the social life of the town, he said: "Of course not—Macun and Shiyan were enemies!"[55]

4   Artisans into Peasants

ONE OF MY INFORMANTS, a former paper trader, asserted in an interview that "four emperors had personally intervened in the Jiajiang paper industry: Kangxi, Qianlong, Jiang Jieshi [Chiang Kai-shek], and Li Xiannian." In fact, only Li Xiannian (vice premier in charge of economic planning in the late 1970s) *personally* intervened in Jiajiang, but it is true that the paper industry has received an un-usual amount of attention from provincial and central governments. One reason for this was that the industry was crucial for the liveli-hood of a large number of people. More important, however, govern-ments understood the centrality of paper for their administrations, for the examination and education systems, and for the enlighten-ment and mobilization of the general population. More than most other goods, paper was also valued as a symbol: it was used in reli-gious ceremonies as an offering, protective charm, or insulating buffer between sacred objects and their profane surroundings. Until the late nineteenth century, written paper was considered sacred and had to be disposed of properly, a task carried out by special charities that collected scrap paper and burned it in a ritually appro-priate way.[1] In elite culture, paper was cherished as the main me-dium of artistic expression and a beautiful object in its own right. In the twentieth century, it came to be seen as a symbol of Chineseness, one of the "four great inventions" (*sida faming*) that constituted an-cient China's main contribution to modern civilization.[2]

Although state interest in rural industries has been a constant from the late Qing to the PRC, the state's approach to such indus-tries has changed dramatically over time. Under what could be called the "old regime of knowledge reproduction," technical control over production lay, with very few exceptions, in the hands of local people. For ideological and fiscal reasons, the Qing state was committed to a rural smallholder economy.[3] According to Susan Mann, Qing officials clung to a romantic notion of the rural economy in which men farmed and women wove in order to fulfill their household needs, without getting heavily involved in the cash economy. Such a division of labor was seen as consistent with moral order and fiscal requirements; other pursuits were seen as disruptive because they induced people to give up farming.[4] In practice, however, Qing emperors and their officials accepted and often actively promoted specialization in cash crops or handicrafts. The central question from the state's point of view was whether production took place within the safe confines of the household: mines and extractive industries were seen as problematic because they attracted large numbers of young men living outside the disciplining bonds of the family; household-based handicrafts of the type practiced in Jiajiang were encouraged because they gave safe employment (from the state's viewpoint) to people who might otherwise became vagrants.

Qing officials and local elites promoted the diffusion of spinning and weaving skills, which were thought essential for a properly gendered social order, but they did not as a rule challenge the monopoly of self-regulating communities over specific production technologies. This restraint was born largely out of necessity: being experiential and primarily nonverbal, knowledge in craft industries was difficult to record and transmit in writing. This began to change around 1900, when new information-processing technologies imported from the West—scientific vocabularies, scale drawings, photography, lithographic printing, and a host of other innovations—made it possible for the first time to take technical knowledge out of the realm of tacit understanding and place it in the hands of literate elites.

The arrival of these new technologies coincided with changes in the self-perception of Chinese elites. From the 1870s on, treaty port publications such as the *Gezhi huibian* (Chinese scientific magazine) introduced Chinese literati to Western technology. From the 1910s on, making (rather than understanding) material objects became a legitimate concern of literati discourse.[5] As urban elites assumed a new role as promoters of modern knowledge, they challenged the legitimacy of traditional monopolies of knowledge. Here, too, treaty port publications set the tone, portraying guilds and other local groups as obstacles to the free flow of information. A much-quoted article from 1886 described how the gold-beaters guild of Ningbo bit to death a member who had taken more apprentices than permitted by the rules.[6] The extensive Republican-era literature on local industries describes guilds and kinship groups as monopolists who hoarded useful knowledge for selfish ends.[7] Progress demanded that knowledge be "made public"—meaning in practice that it should be conveyed in print to literate experts who would manage it for the public good. From the 1920s to the 1960s, urban-based industrial reformers mapped hundreds of rural craft industries all over China and circulated their findings in specialized journals. Although this exercise was framed in the then-current language of technical improvement (*gailiang*), its focus was often more on recording existing knowledge than on disseminating new technologies.

In Sichuan, the transition from the old to the new regime of knowledge reproduction can be dated to the years 1933–36, which saw a total transformation of the provincial power structure. While other parts of China gradually moved toward national unity and developed modern political institutions in the first decades of the Republic, Sichuan stagnated under warlord rule. From 1916 to 1936, Sichuan was *de facto* independent of the central governments in Beijing or Nanjing and ruled by competing warlord factions whose politics owed little to modern ideologies. Although the demand for Sichuan's legal and illegal exports (silk, *tong* oil, hides, pig bristles, and opium) buoyed the provincial economy, warlord hostilities and fragmentation prevented the development of modern industry. On the eve of the Sino-Japanese war, Sichuan's modern sector consisted of a handful of warlord-sponsored factories near Chongqing.[8]

Sichuan was reunited in 1933, when the Chongqing-based warlord Liu Xiang defeated his last remaining rival (and uncle) Liu Wenhui. At the same time, the Communist Fourth Front Army under Xu Xiangqian and Zhang Guotao established itself in the north of the province, and Mao's Red Army, on its Long March to Yan'an, skirted the Chengdu basin. The Communist threat forced Liu Xiang to request military help from the central government in Nanjing.

Jiang Jieshi, who had long hoped to expand Guomindang power into Sichuan, used the Communist presence as a pretext to outmaneuver Liu and bring the province firmly under central control.[9]

The change in political power coincided with an economic crisis, brought about by the world depression and the worst drought in living memory.[10] Regime change and economic crisis combined in the mid-1930s to produce radical change in the province. Within months, the provincial administration was restructured along national lines; tax and currency reforms were initiated; and new factories and arsenals were built. Much of the change was supervised by Jiang himself, who spent a good part of 1935 in Emei county near Jiajiang. Jiang's presence in Sichuan was motivated also by the anticipated outbreak of the Sino-Japanese war: long before 1937, planners in the Reconstruction Committee (Jianshe weiyuanhui) and the National Economic Council (Quanguo jingji weiyuanhui) had determined that Sichuan would be the chief resource base for what was anticipated to be a protracted war against Japan. After the fall of Nanjing in December 1937 and Wuhan in February 1938, Chongqing became China's wartime capital and the main arsenal of "the great hinterland" (*da houfang*). Entire industries were dismantled, shipped up the Yangzi, and reassembled in Chongqing. This sequence of events compressed changes that elsewhere took decades into a few short years. In many ways, Sichuan's warlord governments had continued the late Qing tradition of self-strengthening, with its strong emphasis on local self-governance and provincial autonomy, its predominance of army generals over civilian men, and its absence of mass-based, ideologically motivated parties. By contrast, the Nationalist government that took over in 1935 represents a distinctly modern regime with a strong centralizing and developmental agenda, similar in type to Stalin's Soviet Union, Mussolini's Italy, or indeed to their Communist successors.[11]

Papermakers and the Qing State

Three factors shaped Qing policies with regard to Jiajiang papermaking: a fiscal interest in tribute and taxes, the wish to keep urban markets supplied with paper, and a concern for social order. In 1684, soon after the Qing conquest of Sichuan, the provincial examinations in Chengdu were revived, and Jiajiang papermakers were ordered to supply "paper for the examination sheds" (*wenwei juanzhi*), an obligation that remained in force until the abolition of the examinations in 1905.[12] In itself, the tribute—100,000 sheets of large writing paper, and 10,000 sheets of small *tulian* paper—was not heavy: a one-vat workshop could produce as much in a year. Nonetheless, it provoked litigation between Hexi and Hedong papermakers, which continued to flare up until the Qianlong period. In 1776, the Provincial Administration Commission (*buzheng shi si*, the highest fiscal authority in the province) settled the case. Paper was to be supplied by the "big households" of Hedong, and the "small households" of Hexi were to pay 4.6 *liang* (6.1 ounces) of silver to the "spirit-money association" (*shenfubang*) of Hedong "in order to show that the paper is provided by both parties in compliance with the law."[13] It is not clear how the tribute was levied, but the amount was so small that later sources could claim that "paper production is not taxed." By contrast, paper *traders* had to pay a variety of taxes, the number of which increased rapidly in the late nineteenth and early twentieth century. From the 1850s on, goods transshipped in Jiajiang (including paper) were also subject to a transit tax.[14]

Qing officials did what they could within their limited powers to ensure that paper workshops were supplied with raw materials. Several stelae in the Jiajiang hills report conflicts between owners of bamboo land and poor migrants who collected bamboo to use as fuel or for sale. In order to protect their property, bamboo owners organized "mountain closure associations" (*jinshanhui*), which hired watchmen and organized patrols. County magistrates backed these associations, encouraging landowners to deal with trespassers as they saw fit:

Special order from the county governments of Jiajiang, Emei, and Hongya in Jiading prefecture, Sichuan: . . . According to the petition of local headman Luo, most people in this *bao* . . . depend for their livelihood on bamboo cultivation and papermaking. At the time of the spring harvest, when the bamboo grows and the summer grain is ripening, shameless fellows (*wuchi zhi tu*) take their wives and children to the hills, steal bamboo, and pull out unripe grain under the pretext of cutting grass and firewood. If owners apprehend them, the men accuse them of slandering the innocent, and the women accuse them of rape. People have to suppress their anger

and suffer many indignities. We think it proper to decree a strict closure of
the area and restore order to farming, for the benefit of the people. . . . If
people dare to disrespect written signs posted by the owners and do not listen to shouted warnings but instead wildly slander others, they will be
harshly punished.[15]

Documentary evidence is too scarce to tell if these were instances
of enclosure in which rich papermakers excluded others from mountain land once held in common or a communal defense against vagrants. In either case, the poor were doubly hit, losing both their
customary gleaning rights and their right to a legal hearing in case of
rape or injury. One can easily imagine that local owners read the
stele as an invitation to mistreat "shameless fellows" whose present
and future claims had been dismissed out of hand. Qing support for
the rights of property owners, and disregard for the rights of the
landless poor, was motivated above all by a concern for a properly
gendered, household-based social order. In other parts of the province, papermaking was often carried out by *pengmin* migrants—
groups of young men who cut down bamboo in "waste" mountain
areas and moved on as soon as natural resources were exhausted.
These men, like miners and other unattached male workers, were
seen as a potential threat to the social order. Papermakers in Jiajiang, by contrast, were permanent settlers (although their ancestors
must have arrived as *pengmin*) who worked and lived in stable
households. The 1898 county gazetteer states:

In the Provincial Gazetteer, we read: "Jiajiang honors good manners and
simplicity. The literati are honest, and the people plain." This is still true
today. Farmers, workers, and traders are not extravagant in their daily eating and drinking habits. There is little arable land, and the five grains are
cultivated during three seasons of the year. When there is a break in winter,
people leave home to peddle goods. Bamboo paper and white wax . . . sell
well in many places. People exchange their goods and return, earning
enough for a living. Farmers work hard all year long and do not dare to
pause for a minute. . . . Men till and women weave; they work harder than
people in other counties.[16]

The same emphasis on social stability through properly gendered
work is found in the Republican-era county gazetteer, which expands on the earlier text:

No work is harder than that of papermaking families, and none consists of more different processes. [The old gazetteer] states that "men till and women weave; they work harder than people in other counties." Papermaking surpasses even plowing and weaving: farmers rest between sowing and reaping, workers toil all day but rest at night. Only the families of papermakers work spring and summer, day and night. Old and young, men and women, are all employed in their respective tasks. Local people call this "whole family bustle" (*hejia nao*).[17]

Other Republican-era sources praise the papermakers' frugality, as if frugality in itself was a guarantee of a good social order: "All the people in Jiajiang make paper for a living. They eat coarse grain as a staple and pickled radishes as a side dish. Their life is extremely hard. Because everybody works, crime is rare."[18]

Craft Control and Self-Regulation in the Qing

We find the same positive attitude toward the industry in lawsuits between competing groups of papermakers. Qing-era papermakers formed religious associations known as "spirit money associations" (*shenfubang*) or "Cai Lun associations" (*Cai Lun hui*). On the thirteenth day of the ninth lunar month, each association organized a procession of the god's statue through the township, followed by a banquet in the temple where the statue was housed. In Macun township, the temple meetings continued into the 1940s, although the processions had been scrapped. Meetings took place at the beginning of the main winter molding season and provided wealthy papermakers and traders with an opportunity to discuss sales and loans.[19] Most of the smaller papermakers in the township could not afford to attend the banquet and limited themselves to a short visit to the temple, where they prostrated themselves in front of Cai Lun's effigy.

A stele in the now derelict Gufo (Old Buddha) temple in the hills of Yingjiang township, near Macun, records an 1836 lawsuit between the Cai Lun associations of Yingjiang and several other townships. For unclear reasons—perhaps because the god's statue was housed in Yingjiang—the Yingjiang association claimed pre-eminence over neighboring townships, appointed headmen for their Cai Lun associations, and demanded that they make cash and grain contributions. Headmen who were appointed against their will and refused to serve were beaten by Yingjiang men. The stele text also mentions scuffles, thefts of bamboo, and a case of fraud in the county-level exams, although it is not clear how these issues relate to the central case. What interests me here is not the conflict itself, which cannot be reconstructed from the information on the stele, but the way two successive county magistrates address the "muddle-headed" (*menglong*) litigants:

The use of paper is widespread, and its benefits are general. Generations have made paper for a living and have derived benefit from it. Since you have received your food and clothing from Lord Cai, how could you fail to thank him and worship him as a god? In the first years of the Jiaqing period [1796–1820], your ancestors in this township made a statue of Lord Cai, which they installed in the Gufo temple. In the early years of the Daoguang period [1821–50], they had a wooden statue carved to be carried in procession through the townships and stored in the Guanyin temple after the celebration. . . . You make paper for your livelihood; it is therefore only natural that you have the sincere intention to have an association dedicated to Lord Cai, to thank him for the origin [of your livelihood] and pray for good fortune. Who would be heartless enough to forget the source of his livelihood? In the past, you served [Lord Cai] with utmost dedication. Wherever there was an association, it carried out the ceremonies. The sites were different, but the association was the same.

Although phrased in a language of moral obligations toward previous generations and "Lord Cai," this text can be seen as a statement about property, understood in the sense of "what is proper" to a group of people.[20] Papermaking is the rightful, appropriate occupation for people in the Jiajiang hills because it has been handed down to them by their ancestors, who in turn received the trade from the "first teacher," Cai Lun. This implies certain rights, among them the right to create self-regulating associations, as well as certain obligations, notably the obligation to thank and worship Cai Lun as the "source of their livelihood." As in the kinship discourse discussed in Chapter 2, construing skill as a gift from a distant ancestor or first teacher shifts claims to ownership and control out of the present, where they can be monopolized by individual families or groups, and moves them into a distant past, where they are equally accessible to all.

The magistrate's verdict is preceded by a preamble (*beixu*), written by a local literatus, which puts papermaking in a wider cultural context:

Under the Three Dynasties . . . paper did not yet exist. Big matters were recorded on wooden boards, small matters on bamboo slips. At that time, simplicity was cherished in the affairs of state; therefore, wood and bamboo were sufficient. After the Qin and Han, changes and altercations became more numerous. . . . Accumulated wooden boards and bamboo slips are less convenient than paper. If Heaven had not sent Lord Cai, who with divine intelligence invented paper, any amount of wooden boards and bamboo slips would have been insufficient. Cai Lun was a eunuch of the Han dynasty who reached the rank of attendant gentleman (*shilang*). His biography is included in the *Hanshu* and need not be repeated here.

Papermaking in this vision becomes part of a bureaucratically ordered world: Cai Lun, himself a courtier and administrator, invented paper to ease the work of other bureaucrats struggling to keep up with a growing flood of wood and bamboo documents. By providing the medium necessary for all official communications, papermakers enable the state to fulfill its functions. Although the magistrate scolds papermakers for their litigiousness and obstinacy, there is no doubt that he sees their trade as legitimate and necessary. By emphasizing the orthodox nature of their protective deity, he recognizes the right of papermakers to form self-regulating associations and to articulate their interests.

Changing Modes of Representation in the Early Republic

Cai Lun associations can be understood as an element in what Prasenjit Duara calls the "cultural nexus of power." Leadership in rural society, Duara argues, "was articulated through an institutional framework suffused with shared symbolic values," and it was through the subtle (and sometimes not so subtle) manipulation of cultural symbols, often drawn from popular religion, that local leaders exercised their power.[21] Although open to contestation from all sides, this nexus maintained its integrating power as long as the shared symbols underpinning it were backed up by the imperial state. It shattered in the years after 1900, when the late Qing government and its successors embarked on a series of reforms aimed at the bureaucratization of grass-roots administration, enhanced tax-collecting and policing powers, and the cultural transformation of rural China through the suppression of "superstitious" cults and the extension of modern education.[22] In Jiajiang as elsewhere in China, old modes of integration withered as temples were closed and their land confiscated, and as local elites found new ways to advance their interests in local administration, police work, and education.

The state-building programs of the late Qing and Republican governments were accompanied by attempts to increase revenue and streamline tax collection. These fiscal reforms often had a strongly delegitimizing effect: not only were the projects on which revenue was spent—policing, data collection, and modern education—of little obvious use to rural populations, but the state also failed to impose discipline on the rapidly growing army of official and semi-official tax agents, leading to what Duara has called "state involution"—a situation in which revenue extraction increased not through an orderly expansion of the state's fiscal bureaucracy but through the proliferation of middlemen and brokers who kept a growing share of the revenue for themselves.[23] Sichuan in the early Republican era, one would suspect, was a prime example of state involution. Not only were Sichuan's warlord administrations famously rapacious—some collected the land tax for decades in advance, in one case up to the year 2008—but since the frequent realignments in the power structure also prevented them from establishing a regular bureaucracy, revenue collection often had to rely on extra-bureaucratic agents.

An incident from Huatou illustrates this process. By 1910, every township in Jiajiang had transformed at least one local temple into a junior primary school; by 1935, the number of primary schools housed in former temples had risen to 79.[24] In 1933, the provincial government decreed each township should also establish at least one senior primary school. The township government thereupon confiscated one of the remaining temples in Huatou and began to levy a "vat tax" on papermakers and a turnover tax on traders. The tax revenues far exceeded the needs of a school that never enrolled more than five students. In protest, papermakers revived their dormant Cai Lun association, removed the statue of Hongchuan (a local deity associated with irrigation) from its central place in the temple, and replaced it with a Cai Lun statue that had been stored away in a side room since the late Qing. After weeks of angry demonstrations, county head Luo Guojun investigated the case and petitioned the provincial government on behalf of the paper-makers.[25] In this case, too, papermakers erected a stele to record the official verdict:

We have heard that the prosperity of a country comes from the wealth of its people, and its decline from the poverty of the people. However, wealth and poverty depend on the actions of a small number of men. Greedy people are a lingering poison for society and the state. . . . Since the Qing examination system was abolished, school fees in our township have been low but sufficient. In the last years, the education committee was staffed by venal people who filled their own purses with public money, going so far as to demand taxes for eleven years in advance. . . . Fortunately, last winter county head Luo Guojun came to the township to investigate. Through extensive questioning he came to understand the mood of the people. He petitioned on our behalf for tax exemption and succeeded in having the previous verdict repealed. That we several thousand papermakers and traders retain our livelihood is due only to Sir Luo's Buddha-like virtue.

The Huatou case demonstrates the delegitimizing effects of state-strengthening efforts, but, even more, it shows the enormous staying power of old modes of protest in the paper districts. Thirty years after the Cai Lun cult had lost official backing, Huatou papermakers still rallied around their patron saint and used a rhetoric of livelihood and moral rights harkening back to the imperial era. Papermakers elsewhere in Jiajiang adopted styles of protest more in line with the rhetoric of the new modernizing state, and in this sense, the Huatou protests were exceptional. They were not exceptional, however, in their intensity, nor in the level of success with which papermakers resisted state attempts to increase taxation.

After the abolition of the imperial examination system in 1905, the former paper tribute was replaced by a "paper vat tax" of 5 to 15 percent of output value. In 1917, after widespread protests in Jiajiang, Hongya, and Emei, General Chen Hongfan granted papermakers in these counties freedom from taxation in perpetuity. Reports from the 1920s mention no taxes on the paper industry apart from the transit taxes levied on all goods.[26] In the 1930s, paper traders continued to pay transit taxes; in addition, they had to pay taxes on soda, lime, and bleach, as well as a personal income tax. Together, these taxes amounted to less than 0.13 percent of the industry's output value.[27] Zhong Chongmin's 1943 report lists (1) an education surtax, (2) a tax on chlorine bleach, (3) a business tax, (4) a wartime consumption tax, and (5) an income tax. On paper, the tax load was considerable: the business and wartime consumption tax amounted to 8 percent on value; the income tax to 20 percent on gross income. However, the first two items were farmed out to local tax collectors and collected at tollgates; they amounted to no more than 0.16 percent of the "miscellaneous levies" (zashui), which in turn were only a small fraction of the county's revenue. Items 3 and 4 were based on an assumed price for each load of paper, which in 1943 was 20 percent below the real market price and was further eroded by inflation in subsequent years. Item 5, the income tax, was never systematically collected because there was no way to assess the income of individual traders. Small traders were simply ignored; large ones negotiated lump-sum payments with the fiscal authorities.[28] Former paper traders agree that the only tax that was more or less systematically collected was the transit tax, levied at city gates and river ports, and that even this tax was easy to avoid.[29]

At a time when people joked that "since ancient times there's never been a toilet fee, but nowadays only a fart is free" (zigu weiwen fen you shui, erjin zhiyou pi wu juan), the low level of taxation in the paper industry suggests either unusual restraint on the part of the county government (unlikely, given the venality general among Sichuan administrators) or successful lobbying by representatives of the paper industry.[30] Although, as we have seen, papermakers and their Cai Lun associations continued to protest against taxation under the Republic, the task of representing the industry gradually shifted to paper merchants and their organizations. Paper traders in Jiajiang were customarily grouped into eight groups (bang), named after their final markets. Three of them—the Xi'an, Taiyuan, and Lanzhou bang—were dominated by sojourning merchants from these northern cities, but the members of the Chengdu, Chongqing, Luzhou, Yibin, and Kunming bang were Jiajiang natives. These bang were not formal organizations but simply groups of people who hailed from and traded in the same city.[31]

Apart from the *bang*, there were two paper guilds in Jiajiang: the White Paper Guild (Baizhibang) represented traders; the smaller Red Paper Guild (Hongzhibang) represented dye shops and their workers. The same division between red and white paper guilds existed in Chengdu, where the White Paper Guild (also known as the Jiajiang Guild) represented wholesale paper traders and the Red Paper Guild local paper dyers and retail shops. The White Paper Guild in Jiajiang was relatively unstructured: all paper traders were considered members, whether they had formally joined it or not. The guild did not collect membership fees but raised funds on an ad hoc basis, asking traders to contribute according to their means. Guild leaders were recruited from a handful of wealthy traders, who had enough authority and financial reserves to enforce decisions and negotiate with county magistrates.[32] The Red Paper Guild was smaller but more structured, consisting of a workers' guild (*gongyoubang*) and a masters' guild (*laobanbang*). The workers' guild was open to all paper dyers who had completed the three-year apprenticeship and hosted a banquet for the other members; regular membership fees were not required. The masters' guild required a financial contribution in addition to a lavish banquet for all members. Over the years, the masters' guild amassed substantial capital, which it invested in fields and houses.[33]

In the closing years of the Qing, local authorities were ordered to transform traditional guilds into "same-trade associations" (*tongye gonghui*) and to form overarching chambers of commerce (*shanghui*), intended as consultative bodies that would advise the government and help collect taxes and statistical information.[34] In Jiajiang, these changes remained superficial: despite stipulations that required elected officials and supervisory boards, the new paper trade association (*zhiye tongye gonghui*) continued to operate in the same way as the old paper guilds. Leaders were selected informally from a small group of wealthy traders, and all paper traders in Jiajiang were automatically assumed to be members. The lack of formal structure did not diminish the association's power: one informant recalled that the word of the association's chairman "carried more weight than the order of a court."[35]

The association's success in keeping taxes low probably had as much to do with informal lobbying as with formal representation. We have seen how wealthy papermakers like Shi Ziqing sought the patronage of painters, politicians, and literati. Paper traders were similarly well connected. Take, for example, Peng Shaonong, who managed to combine in his long life (1876–1968) the roles of *xiucai* (the lowest-level degreeholder in the imperial examination system), overseas student in Japan, high-ranking *paoge* member, anti-Qing revolutionary, Guomindang official, journalist, and, late in his life, honored member of the CCP. He rose to prominence as an early member of the Nationalist opposition and, after 1911, served in important provincial positions until he resigned in 1919 out of frustration with warlord politics. He then opened a paper shop in Chengdu but remained politically active as the editor-in-chief of the *Gongshang ribao*, which he turned into a platform for the leftist opposition to the Guomindang.[36] Another prominent merchant was Xie Rongchang, who, as the spokesman of the Jiajiang community in Chengdu, lobbied against wartime proposals to impose price controls on the paper trade.[37] Few of the several hundred paper traders in Chengdu and Chongqing were as well connected as Xie and Peng, but many were linked to intellectual and political circles, if only because this was where their customers were found.

The Grain Crises of 1936–37 and 1941–42

Ironically, the greatest threat to the paper industry came from the land tax, from which most papermakers were exempt. Farmers in Sichuan, whose land tax had been fixed at a low level in the early Qing, paid some of the lowest taxes in Qing China. After 1911, warlord governments tried to make up for low basic tax rates by collecting tax for years and sometimes decades in advance. In Hedong, Liu Wenhui's 24th Army collected tax three to four times a year; in Hexi, six times a year. For most papermakers, who had little arable land, direct tax burdens remained nonetheless low. The problem was the indirect effect of high land taxes: on one hand, increased taxation depressed rural purchasing power and thus the rural demand for paper; on the other, it drove up the market price of grain and thus the cost of making paper. Since grain-based wages and grain consumption in the workshop accounted for about half of production costs, higher grain prices translated directly into higher paper prices. The vulnerability of the paper trade to high grain prices showed during the 1936–37 famine—unprecedented in a province that had been immune to famine under the Qing. Extreme drought in northern Sichuan, combined with intense fighting between the Communist Fourth Route Army and warlord troops, created disastrous grain shortages all across Sichuan. In August 1937, Huang Yonghai, chairman of the paper trade association, petitioned the provincial government:

The harvest of mixed grains was extremely small, and the bamboo harvest decreased by 60 to 70 percent. The price of bamboo more than tripled, and that of rice rose several-fold. By contrast, the price for paper has collapsed. In the last few months, nearly all families have stopped eating rice. The number of those who eat two meals a day of coarse grain has also dwindled. Most survive by eating one meal of maize gruel per day. Since all of Sichuan has been affected by the disaster, the market for paper has shrunk, and paper traders from Jiajiang have one by one given up their trade.[38]

Huang went on to request (1) emergency loans from a disaster relief fund, to prevent a situation in which "papermakers become so destitute that they cannot resume production," (2) technical assistance, (3) an order to all government offices and schools to buy local paper only, (4) the abolition of soda and bleach taxes and the education surtax, and (5) an exemption from the business tax. The petition was addressed not to the provincial government under Liu Xiang, who had been sidelined by Jiang Jieshi, but to Generalissimo Jiang himself. Jiang (or his staff) reacted promptly, issuing an order stating that "all parts of the petition conform to the facts. In order to save more than 100,000 farmers (nongmin) from losing their livelihood, copies of the petition shall be sent to the provincial government, along with an order to draft relief plans."[39] Not surprisingly in a year that saw millions die of famine, the provincial government saw no reason to grant tax reductions to papermakers who still had maize to eat. It did, however, order government offices and schools in Sichuan to stop using imported paper, without specifying how this would be enforced.[40]

The industry recovered quickly, due not to government measures but to a good harvest in 1938. In 1941–42, however, papermakers were once again forced to close their workshops. The reason this time was the central government's decision to collect the land tax in kind, which was in turn caused by the need to feed the army and the urban population, both of which had rapidly increased because of the war.[41] With an average grain quota of 28 kg per household, tax levels in Jiajiang remained relatively low, and many papermakers had so little land that they were not assessed at all.[42] The problem, once again, was that in-kind taxation reduced the amount of market grain, drove up grain prices, and increased production costs. In consequence, paper prices in Chengdu and Chongqing skyrocketed until newspapers and state agencies began to call for price controls or even the nationalization of the entire paper trade. In 1942, magistrate Wang Yunming petitioned the provincial government:

[Jiajiang] is the largest paper-producing district in Sichuan, with an output worth over 100 million *yuan* per year. . . . In good years, the [grain] harvest does not surpass 400,000 market *shi*. . . . Assuming a population of 150,000, yearly consumption is 700,000 market *shi* or more. The grain grown in Jiajiang is thus sufficient to feed the population for only seven to eight months; the deficit must be bought from Hongya and Meishan. . . . During the 1940 purchase of army grain, Jiajiang was recognized as a paper industry district (*zaozhi gongyequ*) and assessed only 3,000 market *shi*, the lowest rate in Meishan prefecture. Based on the former tax rate of 4,031 *liang* of silver, the grain tax for the county has now been fixed at above 81,000 market *shi*. This amounts to 40 percent of the grain income of the landowners, the heaviest burden in all of Sichuan! . . . If this year's taxes will be as high as last year's, the strength of the people will be exhausted, and the paper industry will inevitably collapse.[43]

Paper production, Wang argued, was as essential for the war as farming; since Jiajiang was better equipped to produce paper, it should be allowed to do so. This view was shared by the Guomindang propaganda department, which stressed the "enormous importance of Sichuan paper production for the war of resistance."[44] However, the wartime need for grain was much more pressing than that for paper, and Wang's request was refused. In the end, Jiajiang was once again saved by incomplete enforcement of government policies:

in the next seven years, the county was allowed to accumulate a tax debt of around 1,000 tons, 23 percent of its quota.[45] Corruption at all levels of the tax system also made sure that a large percentage of the extracted grain flowed back to rural markets, albeit at a higher price.

Imperialism and the "Collapse of Chinese Handicrafts"

The grain crises of 1936–37 and 1941–42 foreshadowed the future of the paper industry. Since the Ming-era single-whip tax reform had commuted most corvée duties and taxes in kind into silver payments, the fundamental right of rural people to exchange commodities for cash and cash for grain had never been in question. Now, rural people were redefined as agricultural producers whose main task was to produce grain to feed the nation's soldiers and urban-industrial workers. At the same time, an earlier vision of the economy as a patchwork of interdependent, specialized localities was gradually replaced by the vision of a two-tiered economy, divided into a leading urban-industrial sector and a subordinate countryside. Protests by sympathetic officials that Jiajiang's papermakers were not and had never been peasants came to sound like special pleading, attempts to avoid a burden that all rural people had to bear.

This redefinition of rural people as peasants was accompanied by calls for the reorganization (*gailiang*) of craft industries whose survival was seen as threatened. As Albert Feuerwerker commented long ago, Chinese views on traditional industries were often willfully pessimistic: "The more bankrupt the traditional sector could be shown to be, the more likely it was that a national effort to modernize and industrialize would be undertaken."[46] From the 1920s to the 1940s, numerous reports on papermaking in Jiajiang and elsewhere in Sichuan appeared in provincial journals.[47] In the 1920s, such reports still depicted papermaking as "one of the great handicrafts of Sichuan" and praised it for its contribution to rural prosperity, but by the 1930s, the tone became increasingly negative and calls for reform increasingly strident. This is ironic since the paper industry grew by leaps and bounds in these years. The pessimism was based on the observation that paper imports from Europe, Japan, and coastal China were increasing, and the assumption that other people's gain must be Sichuan's loss. Even observers as astute as Lü

Pingdeng stated that "since the beginning of foreign paper imports, the paper industry of Sichuan has declined."[48]

A compilation of import and export figures does show a growth of paper imports from 6 metric tons in 1891 to 1,116 tons in 1931; it also reveals an increase in paper exports, chiefly to Manchuria, which used large quantities of sacrificial paper money. Throughout the 1920s, the export value was two to four times higher than import value; in 1933, after the Japanese invasion of Manchuria provoked a boycott of Japanese goods, the value of paper exports rose to 24 times the import value.[49] Even this underestimates the real strength of Sichuan's paper exports. Custom statistics covered imports and exports only on large vessels through Chongqing and Wanxian. This meant that practically all imports were covered, whereas handmade paper shipped over land routes or aboard small Chinese craft was often overlooked. Imports of machine-made paper never amounted to more than 5 percent of the output of Sichuan's unmechanized paper workshops. Moreover, handmade paper and imported machine-made paper were used for different purposes, and there is little evidence that the new product was displacing the old. Imported paper was used mainly for lithographic and letterpress printing, which developed rapidly in the Republican period, with more than fifty journals and newspapers published in Chongqing, Chengdu, and other urban centers. Modern-style printing did not immediately replace traditional woodblock printing, which continued to thrive in Chengdu and Yuechi.[50] Rather than being pushed aside, handicraft papermakers appear to have benefited from growing demands created by the introduction of Western print technology. In the 1920s, Jiajiang papermakers learned to make newsprint, and when the boycott of Japanese goods eliminated most of the competition, they quickly became the main suppliers of paper for Sichuan's newspapers and magazines.

Papermakers as Obstacles to Reform

With hindsight, the years 1928–45 can be seen as the Golden Age of manual papermaking in Jiajiang. The end of warlord fighting, first in western Sichuan, then in the entire province, was followed by an enormous expansion in demand for paper after the outbreak of the Sino-Japanese war. By 1943, when the industry was at its peak, an estimated 60,000 persons—one-third of the county's population—derived an income from the paper industry.[51] Yet despite all the evidence for rapid growth, most observers continued to perceive an industry in crisis. Some reports express grudging admiration ("with such traditional production methods, they provide all the paper for the vast hinterland; one can only marvel at the latent strength hidden in our villages!"), yet the main theme in the 1930s–1940s was the backwardness and inevitable decline of craft production. A 1935 report states:

[Jiajiang papermakers] have passed on their trade secrets from father to son for ten or twenty generations. Even friends and relatives who visit the workshops . . . are not told any secrets. This shows the fierceness of their conservatism. Their ignorance of progress stems from the same reason. A government wanting to promote industry must destroy these evil habits (*dapo cixiang louxi*); otherwise it cannot succeed.[52]

Other articles from the period described rural papermakers as conservative (*baoshou*), ignorant (*jianlou*), isolated and unable to cooperate (*lingxing buneng hezuo*), and indifferent to reform (*buqiu gailiang*).[53] The sharpest condemnation was reserved for industries producing ceremonial paper money, called "superstition paper" (*mixinzhi*) in the sources. The use of paper money in ancestral sacrifices had always been considered improper by Confucians, who objected to the expression of filial sentiments in terms of naked cash.[54] Nationalist reformers saw it as yet another proof of the irrationality of China's rural people. In order to force papermakers to shift to the production of "culture paper" (*wenhuazhi*), the provincial Reconstruction Bureau (Jiansheting) slapped an 80 percent tax on "superstition paper." In a document explaining the punitive tax, officials even accused the producers of superstition paper of aiding the Communists:

In the paper districts of Liangshan, Dazhu, Tongliang, Guang'an, Jiajiang, and Hongya counties, large numbers of papermakers have recently produced *huangbiao* paper for superstitious purposes, using national resources for the benefit of the [communist] bandits (*yi guo yuan li fei*). This is an extreme waste of raw materials and labor power. . . . Given the shortage of printing paper, its production must be increased. Superstition paper is a pointless waste of resources, and its production must be stopped.[55]

Another recurrent theme in these documents is the exploitation of ignorant papermakers by unscrupulous merchants. As discussed in the previous chapter, much of the circulating capital in the industry came from merchants who advanced cash or raw materials to workshops, charging interest rates that varied from zero to about 10 percent a month. From the papermakers' point of view, almost any loan was better than a *xianhuo* (cash) sale in an impersonal market, but this was not how outside commentators saw it. Sources from the 1940s describe how merchants took advantage of "farmers without marketing skills," forced by poverty to "walk under the whip of merchant capital."[56] *Yuhuo*, in the view of these observers, "starved the producers and fattened the traders." A 1945 article asks rhetorically: "how many papermakers have been forced out of business during the year? As New Year's approaches, the traders who have tricked the producers through *yuhuo* stretch out their demon claws again. Who knows how many producers will be forced to give up this time?"[57]

Reform from Above

A first attempt to systematically reorganize Jiajiang paper production was made in 1935, after Liu Xiang had made himself the uncontested ruler of Sichuan. Liu's government ordered the magistrates of Jiajiang and seven other counties to reform their paper industries within four years; otherwise they would be fined. Depending on circumstances, counties should establish experimental workshops or centralized factories, for which the provincial government promised funding and technical support.[58] The funding never materialized, despite repeated requests from the Jiajiang paper trade association, and the association itself lacked the means to carry out the reforms. Calls for reform became more urgent with the outbreak of the war and the transfer of the national capital to Chongqing. Paper was now seen as a weapon in the war of resistance against Japan, needed to mobilize the population and sustain morale. Ideally, this paper would have come from mechanized paper mills, some of which had been shipped up the Yangzi from Shanghai at the outbreak of the war. Yet mechanized mills required large investments and advanced engineering expertise, both of which were in short supply during the war. Above all, mechanized mills took several years to become operational, so that even in 1943, after six years of strenuous effort, the mechanized sector accounted for only 20 percent of Sichuan's paper output. Manual production, by contrast, could be expanded quickly and at low cost. Another disadvantage of industrial mills was that they required softwood pulp, not readily available in the heavily deforested Sichuan basin.[59] Bamboo and other grasses (rice straw, sugarcane) were plentiful in Sichuan, growing relatively close to urban areas, where paper was most needed. However, the stiff fibers of bamboo and other grasses cannot be ground to pulp like softwood, nor do they take well to the sulfite process, the cheaper and less complex of the two forms of chemical pulping. China developed industrial-scale technologies for pulping straw and bamboo in the 1950s, but paper quality remained low throughout the 1960s and 1970s. By contrast, handicraft papermakers had perfected the technology for making bamboo paper over many centuries, making paper that was as strong and durable as the best wood-pulp paper. Handicraft papermakers, in short, possessed knowledge of great potential value for the modern paper industry.

The main outcome of Liu Xiang's reform drive was the publication of a number of proposals, some of them written by men with a background in the paper industry, some by self-styled experts. Reform proposals ranged from the cheap and practical (the lining of soaking pits with cement, use of more aggressive chemicals)[60] to the ambitious (replacement of the entire industry with modern paper factories that processed wood pulp instead of bamboo).[61] Almost all these reports portray handicraft technology as inefficient and wasteful, in terms that often betray a lack of familiarity with the local situation. One author, for example, recommends the removal of the knotty bamboo joints before steaming, a proposal that would have required so much labor that it would have priced Jiajiang paper out of the market.[62] Another author proposed replacing fresh bamboo shoots with fully grown bamboo, which he erroneously believes to have more and better fiber.[63] Several reports advocate the replacement of bamboo as a raw material with wood pulp, despite the fact that timber was in desperately short supply.

Technical improvement was only one part of the discussion. Much of the debate focused on the need to replace irrationally small, undercapitalized household workshops with credit and producer co-operatives. Agricultural cooperation—backed in the interwar years by a strong international movement and the League of Nations—enjoyed almost universal political support in Republican China. It was endorsed by the Guomindang as well as by warlord governments in several parts of China, often in versions that owed more to the corporatist models of Fascist Italy or Nazi Germany than to social-democratic or liberal views.[64] Proponents of cooperation in Republican China saw it as a way to break the power of local bullies and merchant elites and extend state power to the village, not as a form of voluntary self-organization, which they opposed on the grounds that rural people were too selfish, shallow, and superstitious to rule themselves.[65] Yet cooperation, like technical reform, required funding. A first proposal to establish papermakers' cooperatives, in 1935, was shelved; it was not until 1945 that a papermakers' cooperative was established with financial support from the Sichuan Provincial Cooperative Funds (Sichuan sheng hezuo jinku).[66] On paper, the cooperative boasted 1,568 members, representing 60 percent of all paper workshops in Jiajiang. In fact, few people in the paper districts knew about the cooperative: it was little more than a scam by the Jiang brothers and other influential men in Jiajiang, who signed people up without their knowledge in order to obtain loans and subsidized raw materials from the Provincial Cooperative Funds.[67]

Toward the end of the war, the argument for cooperation was losing out to proposals to phase out craft industries altogether. In an article discussing the postwar future of Sichuan, the economist Liu Min argued that "if we want to develop Sichuan's industry, we must wage a determined fight against the old model of the self-sufficient economy, against handicrafts and their guild system." The way to develop a modern economy, he continued, was to smash the fetters of the old system, as England, France, and the United States had done in their revolutions. China, however, had to struggle not only against the burden of the past but also against economic imperialism, and traditional crafts could be useful in this struggle. After weighing the pros and cons, he concluded that "the traditional economy, while not the main enemy, remains an enemy and cannot be spared."[68] Another article, dated July 1945, bemoaned the fate of papermakers who suffered exploitation at the hands of greedy merchants but also proposed the construction of a paper factory in Jiajiang that would "completely replace the output of thousands of papermakers."[69]

Reform from Below

Modernizing elites in Sichuan saw industrial reform as consisting of three steps. First, engineers and other experts would go to the paper districts to study and record production methods. Next, they would develop improved methods of production, and finally, these new methods would be returned to the producers, in ways usually left unspecified. The last step would have required a serious extension effort, which was never undertaken—not only because local governments lacked the funds to build model workshops, but also because the experts in charge of the program did not want to leave control over production technology in the hands of local people. In the absence of a viable extension program, industrial reform became in effect a mapping exercise that extracted knowledge from producers and turned it over to an audience of outside experts. It is not surprising that papermakers showed little enthusiasm for what, from their point of view, was little more than industrial espionage. Wealthy and well-connected papermakers like Shi Ziqing occasionally hosted experts from the cities and supplied them with detailed (though not necessarily accurate) information on production costs, turnover, profit rates, and the like.[70] Even such token compliance went too far for many papermakers:

In 1933, the vocational school of Pujiang sent two graduates and asked the county government and reconstruction office of Jiajiang to introduce them to a papermaker so that they could learn to make paper. Given that the request came from the county government . . . the chosen papermaker had to accept it, but after a few days, the rising complaints [from other papermakers] forced him to send them away, making it impossible for the students to study. They asked the county for another introduction, but in order to protect their ancestral secrets and avoid recriminations from their colleagues, one papermaker after the other begged to be excused. In the end, the two students had no choice but to depart.[71]

Although papermakers resented and resisted attempts to expropriate their skills, they were not opposed to technical reform as such.

According to one late Qing source, Jiajiang was one of the first places in Sichuan to import modern papermaking technology: "A paper trader from Jiajiang traveled to Shanghai and Hankou and learned many new papermaking technologies. He collected funds, bought machines and several types of chemicals, and established a factory that produces strong white paper, nearly identical to foreign paper."[72] In 1932, the returned student Su Hanxiang ordered paper machines from Germany.[73] Three years later, the county government had two paper machines "bought from outside the province" in its possession and planned to establish a modern factory.[74] None of these projects left any lasting trace. Change did take place, however, in smaller, less conspicuous ways, as papermakers changed to larger formats, made their paper thicker and stronger, introduced chlorine bleach, and began to use additives such as alum and rosin, which make paper less absorbent. Jiajiang producers learned to make newsprint in the 1920s,[75] and by the 1940s, most of the paper produced in Jiajiang was used for modern-style printing. Although production technology changed relatively little, producers kept pace with rapid changes in market demand. All this took place without government help, almost unnoticed by outside observers.

5   Papermakers on the Socialist Road, 1949 to 1958

IN EARLY 1950, SHORTLY before Chinese New Year's, Shi Dingliang—future party secretary of Macun commune, at that time an unemployed young vatman—went from Shiyan to Jiajiang city. His father had asked him to stay at home until fighting had ended, but Dingliang wanted to buy some meat to celebrate New Year's. When he reached the city, he was stopped by a soldier who called out in a northern dialect: "Stand or I'll shoot! What are you carrying behind your back?" Dingliang dropped to his knees and showed the cloth shoes he had been carrying on his back—like many other villagers, he wore straw sandals and put his cloth shoes on only in town. The soldier answered: "I thought it was a gun. I'm sorry! I'm sorry!" Dingliang returned home without entering the city, shaking with fear and perplexed by this new type of soldier who apologized to a poor boy instead of beating him or dragging him away as a conscript.

After the end of the Sino-Japanese war, the paper industry entered a period of decline as national institutions were transferred back to Nanjing and inflation spiraled out of control. When the People's Liberation Army (PLA) approached Sichuan, terrified merchants went underground, and trade ground to a halt. By late 1949, about nine out of ten papermakers were out of work.[1] The PLA occupied Jiajiang on December 16, 1949, without encountering much resistance. The harvest in Jiajiang had been poor, and the new regime immediately sent a hundred metric tons of relief grain to the paper districts, together with a loan of one million *yuan*[2] and supplies of soda, bleach, and bamboo—all this at a time when both the government and the army were desperately short of grain and pursuing harsh extraction policies elsewhere in Sichuan. Papermakers received preferential treatment because the government needed paper almost as badly as it needed grain. Most paper factories in coastal China had been destroyed in the war, and the outbreak of the Korean War stopped all paper imports into China.[3] At the same time, the government needed large amounts of paper to inform the population about its aims and mobilize people for political campaigns. As delegates from Sichuan's paper industry were told at a conference in Chongqing, "Paper is one of the indispensable weapons [in the movement to resist America and aid Korea]. We have to mobilize the greatest enthusiasm and energy to fulfill this most historical, militant, internationalist glorious production task."[4] In Sichuan, most of this "glorious production task" was of necessity shouldered by handicraft producers. In the Western Sichuan (Chuanxi) district, only 9 percent of the paper output came from modern factories, against 86 percent from Jiajiang workshops and 5 percent from rural workshops elsewhere in Sichuan.[5] For the next several years, the mechanized paper industry proved unable to meet production targets, whereas the handicraft producers easily overfulfilled theirs. The modern factories even used handmade pulp, produced in rural workshops and shipped (in the form of dried bricks) to the cities, because they were unable to produce enough stock for their machines.[6]

Industrial Restructuring and the Socialist State

When the CCP came to power, China's handicraft sector still surpassed modern industry in size. According to official figures for 1952, 6.6 percent of the net domestic product was produced in craft industries, against 9 percent in factories. Handicrafts employed 7.4 million people, versus 5.3 million in modern industry. These figures probably underestimate the real size of the sector: Liu Ta-Chung and Kung-Chia Yeh estimate handicraft employment in 1952 at 13.5 million, against 3.5 million in the factories.[7] Official sources also show a predominance of rural over urban handicrafts: in 1952, 60 percent of all registered artisans worked in the countryside.[8] This number does not include the 12 million "household-based seasonal commodity producers" (a category that refers to relatively specialized rural producers) or the untold millions of rural sideline producers.[9] Yet from very early on, there was a sense that rural craft industries were an anomaly and would eventually have to disappear. Between 1954 and 1956, the number of officially registered rural artisans dropped by over 50 percent from 4.7 to 2.2 million, and most "seasonal commodity producers" were made to join the agricultural workforce. The Great Leap Forward temporarily swept many artisans into the state sector, but when the hastily formed Great Leap "factories" were dissolved, most rural artisans were reassigned not to the handicraft cooperatives of which they had been members but to agricultural cooperatives. By 1964, the number of rural artisans had fallen to less than half a million, down from 10–12 million in 1952.

In CCP usage, "handicraft" signified "a traditional and technically outmoded method of production," whose very existence in the economy was "indicative of a state of economic backwardness which is to be overcome in an historical process of economic development on Socialist premises."[10] Handicrafts were to be gradually transformed into modern mechanized industries, a process that might take several decades and was to be financed by the craft industries themselves. Like farmers, craftspeople were exhorted to "rely on their own strength" rather than expect handouts from the state. The appropriate organizational format for unmechanized handicrafts was the artisanal cooperative; only when cooperatives had accumulated enough capital to mechanize could they hope to be upgraded to state ownership. Despite the influence of artisanal traditions on such urban institutions as the work unit, or *danwei*, the CCP had little sympathy for artisans as a class.[11] Like peasants, they were seen as petty owners of the means of production and thus prone to the "daily, hourly, spontaneous, large-scale reproduction of capitalism and the bourgeoisie."[12] Liu Shaoqi, in a 1953 speech, drew a clear line between artisans and the proletariat: "[Artisans] are not yet part of the working class (*gongren jieji*) and cannot join trade unions. Handicraft co-op members do useful work for the nation and the people; this is glorious. Their social position ought to be clarified: they are laboring people (*laodong renmin*); they can visit the Culture Palace of the

Working People, go to the night school, to the movies."[13] Culture Palace yes, union membership no—small wonder that the party admonished handicraft organizers "not to develop an inferiority complex" (*bu yao you zibeigan*).[14]

The Reorganization of the Paper Industry: Plans and Priorities

Documents from the Ministry of Industry of the Southwest Region (Xi'nanqu gongyebu) and the Industry Bureau of the West Sichuan Administrative Office (Chuanxi xingzheng gongshu gongyeting) show how plans for the paper industry evolved over the first few years of the PRC.[15] They reveal both broad continuities with the Guomindang administration and a much more energetic, hands-on approach. In contrast to its predecessors, the Industry Bureau (later renamed the Handicraft Bureau, and then the Second Light Industry Bureau)[16] had a permanent staff in Jiajiang and was in constant communication with planners in nearby Chengdu. As soon as the hills and mountains of Jiajiang became safe for state representatives to travel, Industry Bureau staff went to the paper districts, interviewed papermakers, and gathered precise information on production technology and ownership structures. By summer 1951, the Industry Bureau had firsthand, up-to-date data on such aspects of the industry as the number of workshops and vats, bamboo acreage, and employment and was able to draft detailed plans for the transformation of the industry.

The grain-paper nexus. From the moment the PLA started bringing relief grain to the paper districts, the exchange of food grain for paper dominated policymaking for the paper industry. The first objective of the Industry Bureau was to receive more paper for less grain; other considerations—technological progress or changes in ownership structure—were always secondary to that aim. This meant that the state had to increase its control over the type of paper produced and the channels through which it was sold. In the late 1940s, many producers in Sichuan had shifted to producing "superstition" paper and cheap writing paper; newsprint had dropped to less than 1 percent of total paper output. From the state's viewpoint, not only were household workshops producing the wrong type of paper, but they were also unwilling to sell to the state agencies set up to purchase their products. Private demand for paper increased rapidly in the early 1950s, and the purchasing agents for schools, colleges, and administrative bodies were going directly to the workshops to buy paper behind the back of state trading agencies. Industry Bureau sources complained that papermakers were "ignor[ing] the needs of society and their duty toward the state" and were reaping "abnormally high profits."[17] This was especially grating in cases in which papermakers had accepted state loans in cash or grain, which in the view of the Industry Bureau created an obligation to sell to state agencies at state-determined prices. In 1953, the private trade in paper was banned, but due to a burgeoning black market, state control remained incomplete. At the same time that paper became a state monopoly, the state introduced the policy of unified purchase and distribution (*tonggou tongxiao*) of grain and banned private grain trade and closed grain markets. On one hand, this increased the state's leverage over the industry, since papermakers became dependents of the State Grain Bureau. On the other, it meant that the state now assumed responsibility for feeding the one-third to one-half of Jiajiang's population that had historically relied on market grain imported from neighboring counties. Such "reverse sales" (*fanxiao*) of grain to rural areas were seen as undesirable, and the authorities in Jiajiang were under constant pressure to reduce the county's dependence on outside grain. The situation in Jiajiang was not, however, exceptional: reflecting the huge proportion of nonagricultural labor in the countryside, almost half the grain collected by the state before 1956 was sold in rural areas, where it supported the cultivation of cash crops, animal husbandry, or handicraft production.[18]

Specialization and collectivization. A second priority, directly linked to the grain-paper nexus, was overcoming the "half-worker half-peasant, dispersed and backward situation" of traditional papermaking.[19] This required the separation of papermakers, who were expected to become increasingly specialized and therefore entitled to state grain supplies, from farmers, who were supposed to be self-sufficient. This was impossible as long as papermaking and agriculture were joined in the rural household: grain intended to boost paper output was being fed to family dependents who were too young, too old, or too unskilled to work in the paper workshops; in some cases, precious state grain was even being used to fatten pigs. Cooperation and collectivization were pursued not so much because they increased labor productivity—since there were no economies of scale in manual papermaking, nothing was gained by grouping several vats together in a collective workshop—but because they tightened the state's control over grain and labor allocation and over paper sales. Collective leaders, it was hoped, would "accept government guidance" and put an end to the "theft of labor and raw materials" (*tougong jianliao*) that characterized household production.[20]

Technological change.   Party leaders were adamant that, with the possible exception of such export-oriented crafts as cloisonné and embroidery, "handicrafts must develop in the direction of semimechanization and mechanization." As Mao said in a speech to cadres in the handicraft administration: "The greater the speed of mechanization, the shorter the life of your handicraft cooperatives. The more your 'kingdom' shrinks, the better it is for our common cause."[21] This might be seen as an injunction for handicraft cooperatives to transform themselves into mechanized factories, but this was not in fact how Mao's remarks were intended. If cooperatives managed to obtain discarded machinery from the state sector or perhaps build machines from scrap, the state might grant them the coveted status of mechanized factory and allow them into the state sector, but they would have to do so entirely on their own. Mechanization was emphatically *not* on the agenda for the Jiajiang paper industry, since this would have diverted scarce investment funds from the state sector and increased competition for raw materials and markets.[22] In the absence of mechanization, little could be done to improve an industry that was already close to the frontiers of what was technologically possible in manual papermaking. It is therefore not surprising that technological change appears in the writings of the Industry Bureau almost as an afterthought. The bureau's experts focused on two issues: increasing fiber yield and shortening the production cycle. The solution to both problems lay in the use of more aggressive chemicals (caustic soda instead of potash) that broke down the bamboo more completely and reduced soaking and steaming time. The trade-off, willingly accepted, was a reduction in paper quality.[23]

Skill transfer.   Another point on the Industry Bureau's agenda was a projected change in the spatial distribution of the paper industry. The Jiajiang industry had long outgrown its resource base: two-thirds of the bamboo and almost all of the grain, potash, and soda it used came from outside the county. From the point of view of state planners, it made more sense to transfer the advanced skills of Jiajiang papermakers to grain and bamboo-rich areas along the borders of Sichuan than to ship bulky grain and bamboo to skill-rich areas like Jiajiang. In contrast to other paper districts, which were scheduled to expand, Jiajiang was therefore to remain the same size for the time being and then ultimately shrink.[24] What state planners overlooked was that craft knowledge could not be transplanted in quite the same way as Soviet technology was transferred to China. As discussed below, skill transfer never worked, despite the evident enthusiasm of the young activists from Jiajiang who went to remote border areas to teach papermaking skills to local people.

Land Revolution in the Paper Districts

Having few local roots, the Communists arrived in Sichuan as military conquerors and ruled initially through the PLA rather than through a civil bureaucracy.[25] For almost a year, the CCP's grip on power in western Sichuan was tenuous, as displaced Guomindang soldiers, *paoge* factions, and local militias fled to the mountains and organized themselves into "anticommunist national salvation armies" (*fangong jiuguo jun*). In Jiajiang, about one-third of these "bandit" forces were papermakers, who had lost their income when the paper industry collapsed.[26] By blocking food supplies and offering pardons to deserters, the PLA slowly reduced the number of resisters, and by early 1951 Hexi was safe for official travel. To consolidate its power, the CCP embarked on a movement to "wipe out bandits and resist local bullies" (*qingfei fanba*). According to a militiaman who participated in the campaign, the rule of thumb was to shoot the "four biggest powerholders" in each *bao* or village.[27] In the tenth *bao* of Macun, Shi Guoliang, son of Shi Ziqing (who had died in 1938), was singled out as a "very big landlord" (*teda dizhu*) and shot. Local informants remember Guoliang as amiable, easygoing, and in no way opposed to the new regime, but like most wealthy papermakers he had been involved in local politics and possessed arms for self-defense.[28] In Huatou, too, the new authorities misread the popular mood in their search for an exemplary villain. They arrested and shot Cao Shixing, the leader of the 1933 revolt against the school tax, who enjoyed as much support in Huatou as Guoliang in Shiyan.[29]

Having defeated or cowed its potential rivals, the CCP began to organize mass organizations in the paper districts. Shi Dingliang recalled how, in early 1950, his friend Zhang Lianming had encouraged him to take part in a meeting to "organize workers" and "form a trade union." Dingliang asked back: "What do you mean by workers and trade union?" Neither term was common in an area where employers spoke of "inviting someone" (*qing ren*) rather than of "hiring" (*guyong*) workers, and wage molders thought of themselves as "selling labor" (*mai gong de*) rather than as workers (*gongren*). Nonetheless, Dingliang joined the union, as did several hundred wage molders and brushers in Macun. In early 1951, Shi Dingliang was asked by his relative Shi Bingcheng to help build a peasant association (*nonghui*). Again, he was puzzled: "Aren't we workers? How can we join the peasant association?" Bingcheng explained to him that "farmers and workers are brothers and have the same aims; it is the same struggle." Dingliang was only half-convinced, but he agreed and became one of the first members of the peasant association in Macun.

Dingliang's bewilderment showed that he had learned his lessons well: brothers or not, farmers and workers belonged to different classes and formed separate organizations. In a 1950 speech, Liu Shaoqi, the party's main spokesman on labor issues, pointed out that the establishment of trade unions outside the cities was "erroneous" (*cuowu*). The revolution in the countryside was to be led by peasant associations; separate meetings of craft workers were permissible but had to remain under the control of the peasant association. In Jiajiang, only Macun—a "test site" (*shidian*) where new strategies were tried out—allowed papermakers to form unions. As Liu had predicted, this policy soon led to problems: union members learned from articles in the *Workers' Daily* that they, as workers, were the true "masters of the country," a status that put them one notch above the members of the newly formed peasant association. An intense rivalry developed between the two organizations, which ended only with the dissolution of the union in 1953.[30]

Although the organization of trade unions suggested that at least some papermakers would attain "worker" status—understood already at that time as more desirable than peasant status—land reform made it clear that *all* rural people, regardless of occupation, were to be considered peasants. Applying Stalin's developmental stages to China, Mao defined Chinese society as "semi-feudal, semi-colonial." In this scheme, feudalism dominated rural China, and colonialism (which simultaneously stimulated and stunted capitalist development) dominated urban China. Since land is the main means of production under feudalism, rural classes were defined in terms of their relation to land.[31] Mao was certainly aware of the fact that his initial breakdown of rural society into five classes (landlords, rich peasants, middle peasants, poor peasants, and landless laborers) did not accurately capture rural realities: in his investigation reports, especially the 1930 *Report from Xunwu*, he freely admitted that many peasants derived large portions of their income from commodity production, small-scale trade, wage labor, and service industries, all of which complicate feudal class relations.[32] Land reform guidelines from the 1950s also allow for the existence of classes that cannot be defined in terms of their relationship to land: paupers, intellectuals, missionaries, vagrants, among others. The guidelines also mention three different categories of artisans: small handicraft producer, handicraft capitalist, and handicraft worker.[33] In practice, however, such classes were often seen as unnecessary complications of the original five-class scheme. This is shown clearly by a proclamation from Huatou township, in which the use of nonagrarian, nonfeudal labels is permitted only in the market town and banned in the rest of the township:

According to the policies and regulations for the assignment of class statuses, we will use the method of "self-appraisal with public discussion, democratic evaluation of decisions, public proclamation of results; the third proclamation is final." Class labels in the township as a whole (*quanxiang*) are: landlord, rich peasant, small rentier, rich tenant, middle peasant, middle tenant, usurer, person living from interest, free professional, poor peasant, farm laborer. Labels for the market town (*jiedao*) are: industrialist-merchant *cum* landlord, industrialist-merchant, merchant, peddler, poor peasant.[34]

Although rooted in Marxist-Leninist ideology, the insistence that class relations in the countryside were almost by definition feudal also served a practical purpose. During the early, "new democratic" stage of the revolution, the CCP actively courted urban professionals, shopkeepers, small entrepreneurs, and other segments of the "national bourgeoisie." In order to prevent the land struggles from spilling over into the towns and cities, where they would damage the fragile economic recovery and alienate these classes, land reform regulations explicitly forbade the confiscation of commercial and industrial property. The interests of local activists and land reform teams were slightly different: in order to generate enthusiasm for land reform and class struggle, they needed to maximize the amount of property that could be expropriated and redistributed. Defining paper workshops as landlord holdings rather than commercial properties fulfilled that purpose, even though it was in violation of stated policies. This sleight of hand was relatively easy in remote Huatou, where local cadres worked without much supervision. In Macun, closer to the county seat, land reform teams had to achieve the same objective in a more orthodox (and thus a more convoluted) way.

Land reform regulations permitted dual classifications: an individual could be given a permanent "family background" label (*jiating chushen* or *jiating jieji chengfen*) that reflected the socioeconomic status of his family at the time of land reform, and a "personal class" label (*benren jieji chengfen*) that reflected his current occupation. Dual classification was standard practice in the cities, where "student of landlord origin" and "worker descended from a small peasant family" were common designations.[35] They were rare in the countryside, where family background and personal class status were assumed to coincide. Dual classification could, however, be used to account for the existence of individual workers, artisans, and capitalists in a society that was by definition feudal. Owners of large workshops were labeled "industrialist-merchants *cum* landlords" (*gongshangyezhe jian dizhu*), which made it possible to "protect them as capitalists while striking them down as landlords." In theory, their "capitalist" property—workshops, tools, and stocks of pulp and paper—was protected, and only their arable land and bamboo forests were to be confiscated. In practice, land reform teams were under pressure to emphasize the "feudal" dimension of these mixed-status households, because only then could their property be confiscated and redistributed, and without redistribution, enthusiasm for the revolution flagged.

Workers, too, could receive dual class labels. Families that derived their main income from wage work were almost uniformly labeled poor peasants; the main wage earners in these families, however, could individually apply for "handicraft worker" status, in which case they (but not their family members) were excluded from the land distribution. Vatmen and other long-term workers were encouraged to apply for this status, perhaps in order to reduce the number of people entitled to shares in the distribution of confiscated property. Apparently, some workers did not understand the consequences of choosing this designation. Former wage worker Shi Rongqing remembers being told that "handicraft worker" status was more "glorious" (*guangrong*) than peasant status, and that only workers with this status would be allowed to hire out for wages. He consequently chose this designation, only to find out that he had signed away his right to receive land.[36]

In Macun township, land reform began in 1951 and was carried out by work teams composed of soldiers, students from other provinces (some informants claimed they were from Guangdong), and activists from neighboring villages. Classification was based on the following rules of thumb: households with a yearly income of more than 40 *dan* (2,400 kg) of grain or employing three or more workers were landlords; those with an income of 20 to 40 *dan* or employing one or two workers were rich peasants; those with an income of 10 to 20 *dan* or frequently hiring short-term labor were middle peasants; those with an income of less than 10 *dan* or living mainly off wage

Table 3
Distribution of Workshops by Class

Township	Landlord	Rich Peasant	Middle Peasant	Poor Peasant	Total	Vats
			Hedong			
Yanjiang	4	14	36	50	104	122
Fuxing	—	1	3	32	36	48
Macun	27	28	150	212	417	453
Zhongxing	26	24	186	149	385	383
Yingjiang	4	16	101	168	289	334
			Hexi			
Yongxing	12	8	59	111	190	175
Mucheng	5	14	66	58	143	125
Nan'an	40	61	462	432	995	925
Huatou	42	89	400	304	835	796
Yuelian	36	87	456	331	910	815
Maliu	23	36	152	186	397	315
Xiema	7	7	10	25	49	52
TOTAL	226	385	2,081	2,058	4,750	4,543
PERCENTAGE	4.7	8.1	43.7	43.5	100	

SOURCE: Sichuan Archives Gongyeting 1951 [13], 1–2.

work were poor peasants.[37] In the Shiyan area, seven out of the 150 to 200 households were classified as landlords and six as rich peasants. The largest owners—Shi Guoliang with eight vats, Shi Haibo with five—were classified as "industrialist-merchants *cum* landlords." The industrial portions of their property were theoretically protected by this label, but both of them were forced to give up most of their workshops, keeping only what they could work with their own household labor.[38] Nonetheless, the total pool of confiscated goods included only ten paper vats, much of the *bao*'s bamboo land, a few houses and sheds, and some furniture, clothing, and bedding confiscated from the "landlords." These were meager rewards for a drawn-out and bitter struggle, and they were more meager still in areas where ownership was less polarized than in Shiyan.

Table 3 shows workshop ownership by class in the entire county on the eve of land reform. In 1951, 4,750 papermaking households operated 4,543 vats: less than one vat per household, because some very small producers shared a single vat. The vast majority (87 percent) of workshops were owned by poor and middle peasants. A better way to look at wealth distribution is to ask how paper *vats* were distributed across classes. According to another 1951 source, landlords in Jiajiang owned 7 percent of the vats, and rich peasants another 12 percent. In other words, the richest 13 percent of papermakers owned 19 percent of vats. In Macun, where ownership was more polarized, the richest 13 percent of papermakers owned 39 percent of the vats. Even this is a relatively flat distribution structure—although we must keep in mind that land reform had a leveling effect even before it started, since wealthy owners sold off or split their property to avoid confiscation.

The conclusion of land reform was celebrated in mass meetings, often culminating in the public execution of landlords and other "class enemies."[39] As in the Republican era, the forms of political violence differed markedly between Hedong and Hexi. Not that Hedong was peaceful: Shi Dingliang recalled that the Macun militia made mortars from hollowed-out trees to bombard bandits in neighboring Zhangyan, and that Yang Xiaocheng—township head and highest *paoge* leader of Zhongxing and vice director of the fraudulent papermakers' cooperative—was executed during a mass meeting, despite loud and persistent protests by his followers. However, such confrontations pale in comparison with a meeting on the banks of the Huatou River, in which 7,000 men and women from four townships commemorated the victims of the old society—of landlord exploitation, Guomindang rule, conscription, corvée labor, famine, banditry, and rape—by erecting a stele (*lingfang*) in their honor. The meeting ended with the public trial and execution of a famous bandit leader.[40]

The Transition to Collective Papermaking, 1952–56

The distribution of "landlord" property resulted in a situation in which most papermakers owned parts of workshops: one-third of a steamer, half a soaking basin, a few drying walls. Led by peasant association activists, workers and poor peasants began to pool their labor and equipment in mutual aid teams (MATs, *huzhuzu*). The first MATs consisted of relatives and neighbors who exchanged labor in the same informal way as they had done in the past. In Shiyan, the

Yuanyi (wide profit) co-op was established by seven poor and middle peasants, all of them Shis of the *ding* and *sheng* generations. New members were readily accepted, and the co-op soon grew to include twenty to thirty skilled workers, several vats, a steamer, and large amounts of bamboo land. In 1953, the first MATs were transformed into handicraft cooperatives (*shougongye hezuoshe*), similar in size and structure to the elementary production cooperatives (*chuji shengchan hezuoshe*) being formed in agriculture at the time. In the handicraft coops, production equipment was jointly owned (although members were allowed to withdraw their shares), and work was directed by an elected leader, aided by an accountant and a cashier. Each co-op had around 150 members, including family dependents. Skilled adults worked full-time in papermaking; unskilled family members tended the fields and helped in the steaming season.[41]

These early co-ops were not territorial units. The Yuanyi co-op, for example, was based in Jiadangqiao (the market street established by Shi Ziqing) and came to absorb most papermakers there. However, those without papermaking skills remained unorganized (no agricultural co-op existed in the area at the time), and those who preferred to work on their own or join a co-op elsewhere were free to do so. This flexible form of organization came to an end with the blanket collectivization of 1956. In agriculture, advanced production cooperatives (APCs, *gaoji shengchan hezuoshe*) replaced the previous patchwork of private producers, MATs, and elementary co-ops. In contrast to their predecessors, APCs were designed to be geographically discrete and "space-filling," in such a way that every rural person and every point on a map belonged to one, and only one, such unit. Membership in this system was determined by place of residence, not occupation or free choice. Blanket coverage and territorial discreteness were desirable in terms of administrative efficiency and social control, but they were difficult to combine with the principle of functional separation between industry and agriculture. Two different solutions were tried in turn: two distinct but geographically overlapping administrations for papermakers and farmers, and a single, dual-purpose administration that included both groups.

In 1956, the 108 handicraft co-ops in the county were merged into 31 large "papermaking production cooperatives" (*zaozhi shengchan hezuoshe*), each of which covered the same area as an APC—typically a village—and included an average of 140 workers.[42] Villages that housed such papermaking co-ops now had two geographically overlapping cooperative units, with separate leaders, budgets, and personnel; in the words of papermakers, industry and agriculture had "split up their household" (*gongnong fen le jia*).[43] Since income from papermaking was much higher than that from agriculture, households were allowed to send only one member—usually their best income earner—into papermaking. One unintended consequence of this was that the line dividing "workers" and "farmers" now ran through villages and families; another was increased gender inequality, since most able-bodied men joined the paper workshops, and the APCs came to include women, children, the elderly, and the unskilled. "Household division" also created problems for administrators, as the papermaking and agricultural co-ops competed for water, land, bamboo, fuel, labor, and managerial talent.[44] Moreover, the village-size papermaking co-ops were large and unwieldy: each enlarged papermaking co-op consisted of four or five workshops, widely dispersed in the hills, and collective leaders found themselves forced to spend most of their time walking from workshop to workshop on narrow mountain paths. After less than a year, in 1957, the enlarged papermaking co-ops were broken up into 60 smaller "combined agro-industrial coops" (*gongnong hunheshe*). The Yuanyi co-op, for example, was cut back from four to two workshops with eighty workers and two to three hundred members.[45] As the name of the combined co-ops indicates, they included skilled papermakers and unskilled workers, who worked in the fields or in the workshops depending on seasonal labor demands. People in Shiyan remember the "combined co-ops" in a far more positive light than the specialized paper co-ops—although they were still too large for their taste.

Leaping into the State Sector

The "combined co-ops" provided one last moment of sanity before the frenzy of the Great Leap Forward. Despite the rhetorical emphasis on small, indigenous technology and balanced development, Great Leap policies were generally hostile to handicraft production. As Carl Riskin has shown, the industries created during the Great Leap "relied in their formative stages upon the annexation of assets of local industry and handicrafts. . . . Redistribution, or 'primitive socialist accumulation,' . . . was the principal resource base for the industrial 'walking on two legs' movement in the Great Leap Forward."[46] In Jiajiang, the sixty "combined co-ops" were merged into sixteen state-owned "factories," with a total employment of 2,500. These were factories in name only; production remained fully manual, as it had always been. The only change was in the treatment of workers: for a brief time, workers in these factories were treated almost like urban-industrial workers, with fixed cash wages of 30 to 40 *yuan* per month, abundant grain rations, extra rations of nutritious food, and rubber boots and protective clothing to shield them from the caustic lye. Like urban workers, they were given one day off each week and promised holidays, old-age pensions, and free health care—none of which ever materialized, due to the short life of the factories.[47] The trade unions, abolished in the paper districts after land reform, were revived. At the same time that a small number of skilled workers were briefly incorporated into the state sector, the remaining papermakers were more firmly integrated in agricultural units. People without papermaking skills and the 80 percent or so of papermakers not recruited into the factories joined the people's communes as part of the general agrarian workforce.

The number of registered papermakers—variously described as the "papermaking," "industrial," or "nonagricultural" population—declined dramatically during collectivization. In 1951, the Industry Bureau reported that Jiajiang's nonagricultural population (*feinongye renkou*) accounted for 51 percent of its total population. Another document from the same year stated that 43 percent of the county's population "depended on papermaking."[48] These numbers included part-time papermakers, household dependents, and persons employed in transportation and related industries. The first detailed count of paper workshops, also undertaken in 1951, reported that 14,006 persons, or 8 percent of the population, were fully and permanently employed in the industry.[49] To cross-check this count, the Industry Bureau calculated the size of the "surplus population," that is, the people in excess of the maximum that local agriculture could sustain; once again, the calculation yielded 8 percent.[50] These 8 percent were seen at the time as the irreducible core of the industry and scheduled for full-time employment in the paper industry. In 1955, the number of full-time specialized papermakers in paper co-ops had dropped to 4.8 percent of the population; under the specialized paper co-ops, it further declined to 2.3 percent. The "combined co-ops" of 1957 brought a brief upswing in numbers, with 4.2 percent of the population registered as papermakers, but under the state paper factories, the number fell again, to 1.4 percent of the population. Not all those who disappeared from the statistics stopped making paper. In Huatou, for example, only one out of four papermaking households joined the specialized papermaking co-ops; all others joined the standard agricultural production co-ops, where they continued to make paper as a "collective sideline."[51] Demand for handmade paper remained strong throughout the collective period, and the industry administration encouraged APCs to produce paper. However, such "sideline co-ops" received grain only in the form of subsistence rations, not the plentiful grain wages given to specialized papermaking units.

Life Under the Cooperatives

Collectivization was perhaps less dramatic for papermakers, who were used to extensive cooperation between households, than for other rural people. For wage workers—the majority of the population in Shiyan—collectivization replaced one set of bosses with another; for small producers, collectivization expanded and intensified customary forms of mutual aid. Work routines remained basically unchanged: work still revolved around vats with six or seven workers, although it was now more common for several vats to be concentrated under a single roof. Piece-rate wages remained the norm, and tasks were still divided by gender, skill, and age as they had been before. Some of the larger private workshops were broken up only for a short period and then reconstituted. Collectivization was thus experienced primarily as a change in ownership, without much impact on day-to-day labor processes. People in Shiyan remember the early 1950s above all as a time of prosperity—certainly if measured against the lean years between 1945 and 1949. Thanks to high paper prices, wages in paper workshops far surpassed those in agriculture. A vatman who made 1,000 sheets of *duifang* every day—an easy workload—could earn a monthly wage of 24 to 30 *yuan*; brushers and pulpers earned 18 to 24 *yuan*, unskilled workers around 20 *yuan*. These were exceptionally high wages at a time when one *yuan* bought six to seven kg of first-grade rice or one kg of pork.[52] Even more important than monetary wages were the state supplies of grain. The Grain Bureau provided the co-ops with 17.5 kg of husked rice a month for vatmen and 15–16 kg for brushers and pulpers. Co-ops set part of these allocations aside for bonus payments and merged the rest with the subsistence rations of the non-papermaking population, to be distributed on a per capita basis. Like urban people, papermakers had to pay for their rations, but prices were so low that people thought of grain as part of their wage, rather than as a commodity.[53] Rice, pork, and liquor had traditionally defined the good life in this part of Sichuan, and in the early 1950s, many papermakers had so much grain that they could eat rice every day, fatten a pig or two, and even distill their own liquor. In the Jiadangqiao market street, teahouses and wine stalls sprung up; neighboring Yangbian, whose residents prided themselves on their sophistication, hired Sichuan opera troupes to perform in the village and invited traveling film-projection teams.[54]

Skill Extraction

Land reform and collectivization in the paper districts redistributed not only land and workshops but also skill and knowledge, as the most experienced workshop owners—who also tended to be the wealthiest—came under immense pressure to share their knowledge with neighbors and representatives of the state. Although supposedly objective and impersonal, land reform punished or rewarded people according to their attitude. One of the few things a wealthy workshop owner could do to improve his lot was to demonstrate his acceptance of the new regime and his eagerness to help his poorer neighbors—for example, by teaching them some of his skills. The principle that expertise could be exchanged for leniency was officially, if belatedly, acknowledged in the case of Shi Guoliang. After Guoliang's execution, the county government reversed the verdict, not because it was considered excessive or unjust but "because a man with such valuable knowledge should be used, not shot."[55]

On the other side of the class divide, the beneficiaries of land reform—wage workers and small papermakers—tended to identify with the aims of the socialist state. Wage workers, in particular, often became activists and leaders and cooperated closely with the Industry Bureau, not least in the transmission of technological know-how. Yet the main explanation for the new regime's greater access to local knowledge lay in the fact that socialism seemed to have removed the need for secrecy. In the past, papermakers thought of their skill and knowledge as an "iron rice bowl": as long as they alone knew how to make a certain type of paper, their survival was secure. Now, the socialist state seemed to offer a much more solid subsistence guarantee in the form of long-term contracts and stable grain supplies. Secrecy and competition seemed to belong to a bygone age: under socialism, everybody's livelihood was secure, and knowledge could circulate freely for the benefit of all.

In 1950, the Jiajiang county government convened a papermakers' assembly (caohu daibiaohui). Delegates were told that the new state would support papermaking with grain, raw materials, and technical advice. In return, all parties involved—workshop owners, traders, workers—were asked to participate in an effort to overcome past inefficiencies. If papermakers wanted the state to supply grain, bleach, and soda, they had to supply precise information on their needs; if they wanted technical advice, they had to allow outside observers into their workshops. Throughout the 1950s, experts from the Industry Bureau (which after the 1953 introduction of market controls became the main administrative sponsor of the industry) recorded detailed descriptions of production methods. The information was then sifted and catalogued; techniques from different workshops and localities were compared, and best practices were identified. Much of this work was carried out in "model workshops," established by provincial order in all paper districts.[56] Once extracted, tested, and recorded, knowledge could be disseminated to a larger audience. The most common medium for this was the survey report (diaocha baogao), written for internal administrative use and characterized by standardized language, careful attention to detail, and exhaustive lists and tables. Technical reports, accompanied by increasingly accurate floor plans and cross sections, were also published in provincial and national industry journals. The technical solutions advocated in these journals appear well adapted to conditions in the paper districts; for the most part, they were cheap (a water-driven stamp mill could be built for 80 *yuan*; improved steamer lids that dramatically reduced steaming time for less than 20 *yuan*)[57] and could be constructed from stone, wood, bamboo, and other local raw materials.

For reasons discussed below, none of the new technologies developed in the model workshops were adopted in Jiajiang. The main direction of technological transmission was from Jiajiang outward, as the Industry Bureau mobilized men and women from Jiajiang to go to remote mountain areas along the border of the province and help the people there to build paper industries. Shi Lanting, who later became president of the women's federation in Shiyan, spent most of the Great Leap Forward, from 1958 to 1961, in Mabian county in the Liangshan Mountains. This is an area with a strong non-Han Chinese (primarily Yi) population, sparsely populated and with abundant bamboo forests. Lanting was one of more than a hundred Macun volunteers who heeded the call to help the "brothers and sisters" in the border areas. She remembered years of hard work in remote mountain villages, but also, in contrast to Jiajiang, sufficient food—she never finished her meals, at a time when people in Jiajiang were starving. The extension policies of the late 1950s were resurrected in the Cultural Revolution decade, another period that put great emphasis on popular mobilization and volunteer work. Wang Yulan was sent to Shaanxi in 1976 to help develop a local paper industry. Like Shi Lanting, she insisted that teaching basic papermaking skills was easy; everybody, she claimed, could learn to brush and mold in a few weeks. The problem, Wang and Shi maintained, was that farm boys and girls could not adjust to the monotony of papermaking and to the relentless discipline required in the workshop.[58] Like other local people, they believed that the capacity for papermaking work needed to be physically imbibed together with the "soil and water" (*shuitu*) of the paper districts.[59] The experience of failed dissemination campaigns seems to bear this out, although we should perhaps substitute "historically grown norms and social structures" for "soil and water."

In Jiajiang itself, papermaking technology remained basically unchanged. One reason for this is that Jiajiang workshops were already close to the limits of what was technologically possible in manual papermaking. None of the innovations introduced to Jiajiang from outside caught on, with the exception of the "hanging mold," a contraption that reduces the strain on the molder's back by hanging the molding frame from the workshop ceiling. Even so, most vatmen found that it constricted their movements and soon switched back to earlier methods.

The Rule of Experts

Like many other rural populations, Jiajiang papermakers were exposed to new combinations of state power and scientific knowledge in the twentieth century. The Chinese socialist state enumerated and classified obsessively, and in so doing created the objects—peasants, workers, landlords—that it purported to describe. State policies in Jiajiang targeted the individual body, whose caloric intake they sought to regulate, and the population, which they divided into occupational and class categories. Activities previously left to individuals or households, such as marketing and the scheduling of work tasks, now came under the purview of state planners. Knowledge previously "buried" in the hands of papermakers was claimed by state-appointed experts in the name of the greater good. This was, in many ways, a familiar twentieth-century story of technocratic elites concentrating knowledge and power in their own hands and in those of an aggressively modernizing state.[60] It is therefore ironic that many of my informants insisted that their life did not change very much in the 1950s, despite revolution, land reform, and collectivization. In fact, many of them were almost indignant when I suggested that the revolution might have transformed their lives. No, I was told, living standards were more or less the same before and after the revolution; no real improvement was felt until the 1980s. Outside political campaigns, social relations remained basically unchanged; changes in gender and generational norms were slow. Most important, everyday routines in the workplace did not change; people still spent most of their waking hours in the workshop, going through the same operations as in the past.

Memories are, of course, reconstructed in the light of the present, and people I interviewed in the late 1990s may have downplayed the momentousness of change. But they point to an important fact. There is something ironic about a modernizing state—a socialist state at that, committed to the view that social life is ultimately determined by the creation and extraction of surplus value at the point of production—that amassed great powers of control and then left the reality of work virtually unchanged. Socialist states have been capable of dramatic transformations in the experience of work,[61] but just as often they have retained capitalist or older forms of organization, or mechanically replicated capitalist forms of management and control. There may be a systemic reason for this: Michael Burawoy and Pavel Krotov argue that state planning strengthens worker control over production, since workers have to fix the problems that arise from erratic supplies under state planning and from uneven or outdated production equipment. Socialist managers, they argue, are often unable to enforce tight quotas and have to cede control to workers in exchange for plan fulfillment.[62]

Several reasons can be given for the paucity of change in the Jiajiang paper workshop, each of them sufficient in itself. The most obvious was the low investment priority given to handicraft workshops, which were told to "rely on their own strength" (zili gengshen)—to transform themselves without substantial input from the state. The model workshops in Jiajiang and elsewhere developed cheap, practical, adapted technologies that could have transformed local papermaking. Indeed, they did transform it—but only after a delay of thirty years, when household workshops enthusiastically adopted small-batch steaming in steel-reinforced pressure steamers, mechanized pulp beating, and other technologies developed or improved by the model workshops. Under the collectives, papermakers knew about these technologies but could not implement them because they required inputs not available to them. Almost all implements in manual papermaking were fashioned from stone, wood, or bamboo; even knives and hatchets were in short supply. Even if pressure steamers and pulp beaters had been within the reach of rural teams, they would have required inputs such as caustic soda, coal, and die- sel oil that were rationed and in short supply.

It is perhaps unfair to fault a development strategy that ultimately achieved its aim of transforming China into an industrial power, even though it did so at great social costs. Given the constraints im- posed by a hostile international environment as well as the desire to achieve industrialization in the shortest possible time, giving priority to heavy and defense industries over artisans and farmers certainly made sense. Yet the scarcity of steel and other inputs was not the main factor in the decision not to develop the Jiajiang industry. More important, it seems, was a rigid formal logic that linked levels of technological development to types of ownership, and types of ownership to status and entitlement. Handicrafts, as Mao had em- phasized, would soon give way to mechanized industry, and artisans would become workers. Both papermakers and state planners under- stood this to mean that mechanization in any form would lead to the upgrading of collective workshops to public ownership and the in- clusion of their workers in the emergent urban rationing regime. From the papermakers' point of view, mechanization was the magic wand that would give them workers' status. Industry Bureau cadres, too, thought of mechanization in terms of entitlement, and therefore as potentially increasing, rather than decreasing, production costs. One of the recurrent concerns in Industry Bureau documents is that even the hope of mechanization could provoke a wild stampede into industry—as indeed happened after 1958.

Industry Bureau cadres were therefore caught in a double bind, told on one hand to transform backward handicraft production into modern industry and warned on the other not to allow handicraft workers into state industry. What they envisioned for Jiajiang was a process of "democratization, rationalization, and enterprization" (minzhuhua, helihua, qiyehua), in which handicraft collectives pulled themselves up by their bootstraps without support from the state—a process of voluntary self-Taylorization, in which papermaking col- lectives increased job differentiation, tightened labor discipline, eliminated slack time, and generally reorganized the labor process in ways that boosted labor productivity. Mechanization would come *after* this restructuring, as a final reward, as it were. This clearly did not work: papermakers were eager to mechanize but unwilling to fund mechanization through increased self-exploitation and deskilling. In the end, the industry remained unmechanized and collective, and papermakers retained a measure of craft control over the labor process. But the price they had to pay for this, as they soon found out, was widespread deindustrialization.

6   The Great Leap Famine and
Rural Deindustrialization

THE FAMINE THAT GRIPPED China from 1959 to 1961 is now widely acknowledged as the worst in human history, with 20 to 30 million deaths in excess of normal mortality; it is also widely agreed that its main cause was not bad weather (although this did play a role) but ill-conceived agrarian policies, combined with heavy grain extraction and a breakdown of the rural distribution system.[1] Sichuan was one of the worst-hit provinces in China, with an excess mortality of 8 to 14 million.[2] Great Leap policies in Jiajiang followed the usual pattern of overconfidence, experimentation, and waste in agriculture and extreme labor mobilization in industry and capital construction.[3] Fields were plowed half a meter deep, smothered in fertilizer (mostly rubble from torn-down houses), and planted twice as densely as before.[4] Crops in these fields developed enormous leaves and stalks but bore no grain. Expecting that "scientific" close planting would result in bumper harvests, communes reduced the acreage sown in grain and gave their members unlimited food in public canteens (*huoshitang*). People were also allowed to help themselves to fruit, biscuits, liquor, and cigarettes at "unstaffed vending stalls" (*wuren shouhuotai*).[5] More than one-third of the population was transferred from the rural sector to fight "pitched production battles" in infrastructural construction and state-owned industry.[6] Each township mobilized thousands of workers to build steel furnaces, which consumed huge quantities of wood and bamboo but produced only useless lumps of iron. One of the infrastructural projects in Jiajiang county, the Great Leap Irrigation Canal, nearly ruptured when filled for the first time; another, the Chafang Reservoir, burst in 1961 and caused major damage in three townships.[7]

Compared to agriculture, the state paper factories and "sideline papermaking collectives" were islands of relative sanity. Factory workers, who were exempted from steel production and capital investment work, continued their work much as they had done in the past. The directors of paper factories "launched sputniks" (exaggerated output figures) with almost total impunity, reporting a daily output of 5,000 sheets per vat, four times as much as the highest realistic figure. Leaders excused this by saying that if they did not inflate the figures, the next-higher administrative level would.[8] The Number Five State Paper Factory in Shiyan reported fictive workers, for whom it collected wages and grain rations, and overdrew its bank account, which it could do because credit limits were linked to reported output value. Overdrafts and excess wages were distributed to the factory staff and workers. As one former accountant explained, such practices were common and carried little risk: "Investigate? Investigate my ass! At that time, nobody would find out anything" (*Cha? cha ge pi! Dangshi cha bu chulai*).[9]

Jiajiang's grain harvest in 1959 was 15 percent below that of the previous year, mainly because the area sown in grain had been reduced. At the same time, per capita procurement rose from 103 kg of husked grain in 1958 to 136 kg in 1960, as huge quantities of grain were shipped from Sichuan to other provinces and to the Soviet Union.[10] Average grain availability fell from 254 kg per capita in 1958 to 121 kg in 1961. Death rates increased from 13.9 per thousand in 1958 to 26.3 in 1959 and then to a staggering 102.6 per thousand—one out of ten—in 1960.[11] At its peak, the county's death rate was almost twice as high as the provincial average, at a time when Sichuan had the second highest mortality in China.[12] The grain distribution system that had supplied the paper districts was one of the first casualties of the Leap, as transfers from the plains and from neighboring counties broke down. The 2,500 employees of the state paper "factories" were largely shielded from the famine: like other employees in the state sector, they continued to receive rations, albeit much reduced. However, the vast majority of papermakers had been merged back into the general agricultural workforce, and like other

rural people, they were left to fend for themselves. When grain supplies broke down, starvation followed: in 1959, people who still had
rice chaff to eat counted themselves lucky; in 1960, people ground
maize cobs and tree bark into flour. Collective units in Macun introduced a strict distribution scale: infants received 100 g of un-
husked rice per day, children 200 g, the elderly 300 g, working
women (classified as "half-labor") 400 g, and working men 500 g per
day. Some grain was reserved for people suffering from hunger edema,
a measure that saved many lives.[13] The first to die were often former
landlords and rich peasants, who had been put under the "supervision of the masses" and therefore did not dare to glean unripe grain
from the fields or dig out roots in the hills. One man whose bad-class
father and grandfather had died early in the famine remarked bit-
terly that "cadres did not starve." This was not universally true: a
leader in the No. 2 production team in Shiyan was so scrupulously
honest that he was the first in his team to die.[14] At the end of the
famine, some townships in the hills had lost 40 percent of their
population.

Paper production continued until people became too weak to
work. In 1960–61, discipline broke down; workshops were plundered
and everything that could be sold was carried to the plains and exchanged for food. In the same year, papermakers started to cut down
their bamboo and sell it to farmers in the plains. One of the many
ironies of the Great Leap Forward was that starving papermakers
were mobilized to fight a plague of bamboo locusts, at a time when
the famine made them destroy more bamboo than the locusts ever
did. In 1960, after four years of anti-locust campaigns in which
women, men, and children searched the hills for locust eggs, victory
was declared. By that time, papermakers had rooted out large tracts
of bamboo forest and planted maize or sweet potatoes on the slopes.

Until the post-Leap famine, the papermaking hills of Jiajiang
were almost as densely populated as the agricultural plains. The average population density in the ten townships where papermaking
was most concentrated was 242 persons per km^2, against 273 persons
per km^2 in the agricultural townships. Macun (275/km^2), Zhongxing
(312/km^2), and Longtuo (354/km^2) were as densely populated as some
of the most prosperous townships in the fertile, irrigated plains

around the county seat. When famine struck, it hit the paper districts much harder than the plains. The papermaking townships of western Jiajiang lost one quarter (24 percent) of their population between the last pre-famine and the first post-famine population counts, whereas the agricultural townships lost only 4 percent.[15] Population losses correlate to some extent with distance from the county seat and were particularly high in the remote and inaccessible townships of Huatou and Maliu. Yet Nan'an, only a few hours' walk from the county seat, also lost one quarter of its population. A much stronger correlation exists with the size of the papermaking population. Huatou and Maliu, where 43 and 34 percent, respectively, of the population in 1951 depended on papermaking, registered population losses of 40 percent; Longtuo and Nan'an, where 47 percent of the population used to depend on paper, lost 31 and 26 percent; Zhongxing (28 percent) lost 22 percent.[16] As China's demographic expansion in the 1960s demonstrates, famine-struck populations can recover very quickly, even after dramatic population losses. What is striking about the Jiajiang hills is that a full recovery did not take place until the 1990s. Whereas populations in the agricultural parts of Jiajiang soon exceeded the pre-famine levels, most townships in the hills saw little growth between the 1950s and the 1990 census. Taken together, the papermaking townships had a population increase of 35 percent in more than thirty years—one-half of the 70 percent increase in eastern Jiajiang. Individual townships experienced even less: in Huatou, the population increased only 19 percent between 1951 and 1990; in Mucheng, 9 percent; in Longtuo, 7 percent—clear evidence of long-term poverty.

Recovery . . .

In 1961, Shi Dingliang (the former trade union and peasant association activist, who had risen to the post of party secretary of Macun commune) and several other commune leaders wrote a petition to the central government, in which they demanded to be punished for their mismanagement of the paper industry. Whoever was to blame for the collapse of papermaking, it was not the leaders of the commune, who had lost control over the workshops when they were transformed into state-owned "factories." But self-accusation was the only acceptable way to draw attention to their problems; pointing fingers at anybody else would have raised the suspicion that they were opposed to provincial and perhaps central policies. It was a desperate measure, taken only after all members of the party cell of Macun commune had appealed in vain to the county, district, and provincial governments. The county government had expressed sympathy, but "it had no grain power"; the provincial government under hardliner Li Jingquan had ignored their letters. Two weeks after they had sent the letter to Beijing, an unannounced motorcade with officials from Beijing, Chengdu, and Leshan arrived in Macun. The first thing Dingliang did was to chase away onlookers: "Many of them had swollen bellies [caused by hunger edema], and we did not want the leaders to think that we were organizing a demonstration." During the meeting that followed, Dingliang argued his case against Industry Bureau cadres, who accused him of exaggeration. In the end, Dingliang was asked to draft a rescue plan. He concentrated on three points: grain supplies to papermakers should be raised to 19.5 kg for full workers, paper prices should be raised to around 150 *yuan* per 10,000 sheets, and papermakers should be allowed private plots (*ziliudi*) of one *fen* (66.7 square meters) per capita.[17] These suggestions were accepted; after a test run in Macun, they were extended to all paper districts in Jiajiang, Hongya, and Emei.[18]

Dingliang's rescue plan coincided with central policies that aimed to revive the collapsing economy and rectify some of the Great Leap excesses. The Sixty Articles on the Communes stipulated that communes should "generally not run new enterprises for years to come. All the enterprises already undertaken, which are not qualified for normal production and are not welcomed by the masses, should be stopped without exception [and] be transferred to the handicraft co-operatives and production teams for management or changed into individual handicrafts."[19] It also specified that handicrafts in the countryside had to "directly serve agricultural production or the life of peasants." The lesser-known Thirty-Five Articles on Handicrafts, promulgated in 1961, ordered that state and commune industries formed on the basis of former craft industries be disbanded and their assets returned to lower levels. Since "handicraft workers under the rural people's communes have very close links with agriculture, with the exception of the few places where handicraft is particularly concentrated, in general no [specialized] handicraft cooperatives will be established." The main form of handicraft production in the countryside was to be the "small handicraft production group," owned and managed by the brigade (*shengchan dadui*) or production team (*shengchan dui*) and staffed by "half-worker, half-peasants" who divided their time between handicraft production and agriculture.[20] Later regulations specified that specialized handicraft cooperatives should be concentrated in county seats and market towns. In the villages, the standard model was to be "half-worker half-peasant handicraft production" under the production teams.[21]

Although decentralization of the large, inefficient Great Leap industries was clearly necessary, these regulations did not just restore the pre–Great Leap situation but went far beyond it. With the exception of small cooperatives of basket weavers, toolmakers, and the like in market towns and of individual village artisans, *all* rural artisans were to be merged back into the agricultural workforce. Commodity production for urban markets, so common across rural China until the 1950s, was not banned—papermaking continued at a much reduced level, as did many other craft industries—but it lost its legitimacy. In Jiajiang, paper workshops came under team management when the "state paper factories" were disbanded.[22] Initially, people welcomed the change, since they thought of the post-Leap agricultural production teams as a continuation of the pre-Leap "combined co-ops," in which agriculture and industry were flexibly combined. The reorganization was accompanied by grain and capital injections that helped to revive the collapsed industry. The Second Light Industry Bureau (the new name of the Handicraft Bureau), which had been in charge of papermaking during the Great Leap, was made to pay for the losses incurred by the state "factories." The Number Five Paper Factory alone had appropriated 330,000 *yuan* worth of equipment from the paper co-ops; after the Great Leap, less then one-tenth of the capital was left. Workshops had been demolished and bamboo groves cut down, paper walls had crumbled, paper pulp had rotted in the vats. Not all losses were restored, but the provincial Light Industry Department offered 120,000 *yuan* of interest-free loans to papermaking teams.[23] More important, papermakers were once again given generous grain rations. The special grain provisions for "industrial" populations in the paper districts had been abolished in 1961. Instead, the county Grain Bureau provided subsistence rations for everybody in the paper districts, and the Light Industry Bureau introduced output-related grain premiums (*jiangshouliang*) of 100 kg of grain for every 10,000 sheets of first-grade *duifang* and 75 for second- and 50 kg for third-grade *duifang* paper. In addition, teams were given cotton-cloth premiums for high-quality paper. This translated into an extra daily grain wage of 0.5–1.0 kg for each worker, enough to make papermaking attractive again. Thanks to increased rations and sales prices, papermaking soon recovered. For three years, from 1962 to 1965, papermakers had once again enough grain for themselves, enough indeed to feed some of their rice to the pigs.

During the famine years, papermaking in many parts of Jiajiang had slipped out of collective control. Collective order had completely broken down after 1958; what was euphemistically called "canteen decentralization" (*huoshitang xiafang*) was in fact an almost total devolution of production and consumption to the households.[24] In Hexi, most people had never experienced functioning collectives: collectivization arrived there only in 1955, followed by hasty reorganization and then by famine. In Hedong, the collective order survived in better shape, but even there, many teams had reverted to household production, thinly disguised as "contracting output to the household" (*baochan daohu*) or "to small production teams" (*baochan daozu*). In 1962, 41 percent of all agricultural production teams in Jiajiang practiced household production.[25] Under the slogan "land is collective, bamboo is collective, but we each live our own lives" (*tudi shi jitide, zhulin shi jitide, shenghuo ge guan gede*), teams rented out their workshops to households or groups of households, charging them management fees but otherwise leaving them alone. Light Industry Bureau cadres were aware of these practices: as Xie Baoqing, former director of the bureau, admitted, some 80–90 percent of papermaking teams in Hexi practiced household or small-team production, but the majority of Hedong workshops remained under collective control.[26] At the same time, the paper trade was slipping out of the hands of state traders. As the urban economy recovered, urban shops and work units sent purchasing agents (*caigouyuan*) to Jiajiang to buy paper directly from the producers, and representatives of teams and brigades in the paper districts started to sell paper on the black and gray markets.[27] In 1962, state trading companies purchased only 197 metric tons of paper, the lowest figure ever; in contrast, an estimated 200–300 tons was bought by black and gray market traders, who paid 30 percent more for each sheet of paper than the state buyers.[28] Grain premiums must be understood in this context: not only as an aid to help the industry recover but also as part of the state's attempt to regain control over output.

. . . and Renewed Decline

Life in the paper districts deteriorated after 1965 as quickly as it had improved in 1962. The key issue, once again, was grain. So-called reverse sales (*fanxiao*) to rural areas, common before the Great Leap Forward, were abandoned after 1965 in favor of a policy of local self-sufficiency. The impetus, it seems, came from Mao Zedong himself, who believed that not only the nation as a whole but every locality from the province down to the commune should strive for maximum self-sufficiency in grain and other products.[29] Agricultural policy after the Great Leap is often described as a "grain first" policy but, as Yang Dali has shown, the percentage of land sown in grain actually declined in the post-Leap years. Chastised by the famine, the state reduced its share of the harvest from close to 40 percent in 1959 to around 20 percent in the 1970s. Reduced grain extraction made it possible for grain surplus areas to take some of their land out of grain cultivation and plant oilseeds, sugarcane, and other cash crops. By contrast, areas that had historically specialized in cash crops and handicraft production lost their access to market grain and were forced to become self-sufficient. The overall trend was toward reduced specialization and reduced interregional exchange.[30]

In 1953, 33 percent of the county's households were classified as "unified purchase households" (*tonggouhu*), that is, households producing surplus grain that could be purchased by the state. Twenty-nine percent of households were classified as basically self-sufficient, and 38 percent as grain-short and dependent on state grain supplies (*tongxiaohu*).[31] In 1959, the county was ordered to achieve grain self-sufficiency, and county borders were adjusted to make it easier to achieve that goal: the papermaking township of Yuelian was transferred to Emei county, and five grain-surplus townships to the north and east were added to Jiajiang. Thereafter, Jiajiang could no longer count on grain transfers from neighboring counties; if the county wanted to continue to produce paper, it had to distribute grain from its own farmers in the plains to papermakers in the hills.[32] Although the plains of Jiajiang began to produce substantial grain surpluses in the 1960s, the political and economic cost of extracting grain had become prohibitively high. State grain procurement fell from 31 percent of the harvest in 1962 to under 20 percent in the 1970s.[33] With reduced grain extraction and a growing urban population in the county seat and market towns, little was left for redistribution. Grain subsidies for papermakers were increasingly criticized as wasteful: after all, the argument went, these people were peasants and should be able to feed themselves.[34]

Among the first casualties of the push toward self-sufficiency were the grain premiums, which were abolished in late 1965. At the same time, the semi-legal rural markets that had sprung up in Jiajiang were closed, and black market trading was suppressed.[35] Subsistence rations to papermakers were greatly reduced and, in most parts of Jiajiang, eventually abolished. These rations were based on a monthly consumption quota of 14 kg of husked rice per capita, with fixed conversion rates for wheat and "lesser" grains.[36] Local people say that 14 kg of rice was too little to sustain vatmen and pulpers, especially when other food was scarce and almost all caloric intake came from grain. However, since children and the elderly consumed less than adult workers, a per capita average of 14 kg was considered adequate. In fact, the State Grain Bureau never supplied the full 14-kg ration; rather, it allocated only the difference between assumed consumption needs and assumed local production. Both figures were calculated at the level of the production team, on the basis of population counts and projected harvests.

Until 1965, many teams still received half to three-quarters of their grain—conventionally expressed in so and so many months of state supplies—from the county Grain Bureau. After that date, strong pressure was brought to bear on communes, brigades, and teams to achieve total or near-total self-sufficiency. Most teams in Bishan village (near Shiyan), for example, received six months' worth of state supplies, which allowed half the population to work in collective paper workshops. In 1965, these teams were declared self-sufficient, and state grain supplies were stopped.[37] Faced with a sudden halving of their food supply, people cut down bamboo and began to plant food crops in the hills.

Pressure on teams to become self-sufficient continued throughout the 1960s and 1970s, in a variety of ways. Most commonly, teams were given unrealistically high grain production quotas; protesters were told that they could easily achieve the target by opening up "wasteland" in the hills. At the same time, consumption quotas were first frozen and then gradually reduced. After 1968, new team members (children and married-in wives) were no longer added to the list of grain recipients, with the result that per capita consumption fell from 14 to around 12.5 kg in Shiyan. Moreover, rice, the only staple food considered palatable by most rural Sichuanese, was increasingly replaced by wheat, maize, and—considered worst of all—sweet potatoes. A further blow came in 1976, when it was ruled that only "paper peasant teams" (teams that derived more than 70 percent of their collective income from papermaking) were entitled to grain subsidies.[38] In consequence, paper production became increasingly concentrated in Macun and Zhongxing township in Hedong, and in Nan'an, Longtuo, and Huatou in Hexi. No count of people still engaged in papermaking is available, although the drop in paper output from 2,739 tons in 1965 to 511 tons in 1977 suggests a corresponding drop in employment from around 5,000–6,000 in the mid-1960s to 1,000 in the mid-1970s.

The tightening of grain supplies was accompanied by increasingly strident campaigns against moneymaking tendencies in the countryside. Papermakers who received state grain were accused of eating "guilty conscience grain" (*chi kuixin liang*) and of "being half-assed" (*pigu zuo wai*), meaning that they, as peasants, should focus all their efforts on agriculture. These attacks took place in the context of an effort to "trim and simplify" the urban population (*jingjian chengzhen renkou*) and the workforce in the state-owned industrial sector, both of which had grown dramatically during the Great Leap. The aim (almost achieved) was to reduce the urban population from 120 to 100 million, and the nonagricultural workforce from 40 to 30 million, by moving recent rural-to-urban migrants out of the cities and recently hired state employees into collective handicrafts and agriculture. Although the work teams that carried out these policies were explicitly told that this was "not a political campaign" and that there were to be "no struggle meetings, no wearing of hats, pulling of braids, or beating with sticks," "trimming and simplifying" quickly morphed into an aggressive campaign that eventually merged with the "Five Anti" and "Four Clean-up" movements against capitalist tendencies and corrupt or ideologically lax cadres.[39] Documents from the mid-1960s reveal a wide range of tactics employed by handicraft co-ops: various forms of subcontracting, underreporting of profits, paying of "excessive" dividends and wages to members, and sale of state-supplied materials and finished products on the black market, among others.[40] They also show the growing exasperation of the state leadership with recalcitrant handicraft producers. The CCP had always considered artisans a hybrid class, torn between the selfish instincts of small property owners and the socialist inclinations of proletarian workers. The stubborn resistance of rural craft producers to increased state control clearly demonstrated that handicrafts, because of their "local and dispersed nature," were particularly prone to "splittism, localism, bureaucratism . . . individualist and capitalist tendencies," which in their extreme forms constituted "criminal activities aimed at a capitalist restoration."[41]

Demand for Handmade Paper

Was the decline of manual papermaking a painful but necessary outcome of the growth of mechanized papermaking? In 1957, handmade paper still accounted for a quarter of China's total paper output, but as the modern paper industry expanded, machine paper replaced handmade for many usages.[42] Market demand is a tricky concept in an economy in which consumption, like production, is in theory state-planned. How much and what kind of paper was consumed in China was very much a matter of state planning and politics. In the 1960s, 105 paper mills all over China were pressed into service to provide paper for 800 million copies of Mao Zedong's works, and paper production for other purposes was put on hold.[43] On a smaller scale, stockpiles of 1,200 tons of Jiajiang paper were sold out within weeks after the outbreak of the Cultural Revolution, because students needed paper to write "big character posters" (*dazibao*).[44] Outside political campaigns, state planners gave priority to newsprint, paper board, nonabsorbent writing paper, and packaging paper, all of which were produced in growing quantities in the modern sector.[45] Other demands may have existed latently but were not recognized by state planners. Handmade paper was traditionally used for a wide range of purposes: for calligraphy and art, interior decoration, and most important for popular religion, where in addition to the sacrificial burning of paper money and other objects, paper was used for scrolls, streamers, firecrackers, and so on. Almost all these usages came to be associated with bourgeois high culture or feudal superstition and were either ignored or suppressed.

In the Maoist years, annual per capita paper consumption in China was one of the lowest in the world: 4.5 kg per capita in 1968, against 242 kg in the United States and 24 kg in the Soviet Union.[46] Most of this already small amount went to the cities. Judging from anecdotal evidence, rural China in these years was an almost paperless society in which pupils in village schools wrote on wooden boards and toilet paper was considered a luxury. In the 1950s, Jiajiang produced mainly writing and printing paper for urban markets. In the 1960s, machine paper began to replace handmade paper in this section of the market, and Jiajiang papermakers changed to producing *duifang*, a medium-size, medium-quality paper that can be used for writing, printing, and other purposes. Production of high-quality calligraphy paper declined. Although *duifang* was often termed "culture paper" in official documents, implying that it was used as writing paper, it was an open secret that most *duifang* ended up in smoke, being burned at gravesides and on ancestral altars. About a quarter of Jiajiang's paper output between 1950 and 1985 was exported to Southeast Asia, especially Hong Kong, Burma, and Singapore, where it was used as sacrificial paper. According to Xiao Zhicheng, former head of the Light Industry Bureau, Jiajiang paper was sold at such low prices that paper factories abroad bought it as raw material and pulped it.[47]

Domestic sales were largely in the hands of the Supply and Marketing Cooperative and its affiliated trade firms. Although the demand for Jiajiang paper always exceeded supply, state pricing practices could lead to stockpiles. This happened in the mid-1960s, when purchase prices were raised to revive production, but sales prices in the urban market remained fixed. This led to "inverse pricing" (*jiage daogua*)—a situation in which paper could only be sold at a loss—and in consequence to growing stockpiles. This problem was solved when Yao Yilin, minister of commerce and one of the PRC's foremost economic planners, visited Jiajiang and promised to subsidize the losses of state trading firms.

Digging out the Bamboo Roots

Unplanned bamboo cutting in the famine years soon gave way to systematic deforestation, as teams came under pressure to increase grain production. "Wasteland" in Jiajiang was bamboo land; before planting maize and sweet potatoes on the slopes, people had to cut down the bamboo and slowly, laboriously, hack through the tangled bamboo roots. Between 1958 and 1975, nearly half of Jiajiang's bamboo acreage was destroyed. Without the rhizomes that held the loose red soil together, the hills became prone to erosion during summer rains. In Longtuo township, for example:

After more than 10,000 *mu* of bamboo and forest were cut down, maize harvests increased slightly, but bamboo was depleted and papermaking was deprived of raw materials. The economic losses were huge; Longtuo became a grain-short, cash-short impoverished mountain district. In the following years, storms, hail, and floods followed one another. Landslides destroyed streets and bridges and buried the headquarters of the township government and the Supply and Marketing Cooperative.[48]

Wherever teams expanded agriculture at the expense of papermaking, incomes fell dramatically. Nominal incomes, calculated in work points, remained higher than in agriculture: vatmen in Shiyan earned 20–22 work points for every 1,000 sheets of *duifang* (a day's work for a strong man), and brushers and pulpers about 15 points a day. In contrast, most agricultural work earned 10 points a day. However, with declining income from paper sales, the real value of a vatman's workday fell from about one *yuan* in the 1950s to about 0.73 *yuan* in the 1970s.[49] Even more important were dietary changes. Rice had always been the staple food of papermakers; corn and sweet potatoes were eaten as supplements in times of need. After 1966, state-supplied rice became increasingly scarce and had to be supplemented with home-grown corn and sweet potatoes. This was not simply unpleasant; in a culture where the food one eats is a main indicator of one's social standing, it was socially degrading.[50] Because of the norm of reciprocity in banqueting and gift-giving, eaters of corn and sweet potatoes found it difficult to keep up social intercourse with rice eaters in the plains. Even more important, food was a main consideration in marriage decisions: parents from the plains were unwilling to marry their daughters to families in the hills, where they would have to subsist on coarse grain.

Not all changes can be quantified, and those that cannot are not the least important ones. The paper trade had served as a bridge between the Jiajiang hills and the urban centers, bringing protection, patronage, technical information, news, and gossip to the hills. With the decline of papermaking, the Jiajiang hills became a periphery, described by its inhabitants as "desolate" and "remote" (*pianpi*)—terms that would have made little sense in earlier periods, when dense economic and cultural ties linked the paper districts to urban centers.[51]

Life Under the Teams

Where papermaking continued through the 1960s and 1970s, work procedures remained largely unchanged. Quite apart from the shortage of inputs such as steel and iron, production teams (since 1963 the formal owners of the workshops) had few incentives to improve technology. What was the point in producing more paper, or producing it more efficiently, if more paper did not buy more grain, at least not in any predictable way? If frequent changes in state policies had taught papermakers anything, it was that the amount of grain in their bowls depended on administrative fiats coming down from Chengdu or Beijing, rather than on how hard or efficiently they worked. Another reason not to change production technology was persistent underemployment in the teams. Team leaders were hard pressed to find sufficient jobs for a steadily expanding workforce, at a time when workshops could not easily expand. Any changes that would have improved labor productivity would have reduced the number of available jobs and forced more people into low-paying agriculture.[52]

One of the few positive changes in the workplace was the reorganization of women's work. Although woman workers had never been confined, in a strict sense, to their houses, they spent much of their workday literally staring at the walls of their homes and courtyards while pasting paper, usually in isolation. After 1956, drying walls came under collective management. Most of them were still on the inner and outer walls of people's houses, but they were maintained by the agricultural production team and used by roving teams of women who went from house to house, pasting paper wherever they found free space.[53] Working in teams reduced the tedium of brushing, and women experienced the increased mobility as liberating. The income of skilled brushers remained lower than that of vatmen but higher than that of other male workers in the industry. During the 1960s, when women were told that "everything that men can do, women can also do," a few women even learned to mold and make pulp, although to my knowledge none of them became a fully specialized paper molder or pulper.[54]

The events of the Cultural Revolution had little impact in an area with few schools and few students. Shi Shengyan, son of a Macun papermaker, became the leader of one of Chengdu's Red Guard factions, but this was long after he had left Jiajiang. Middle school students in Macun joined the Red Guards if they were of poor or lower middle peasant (pinxiazhongnong) descent, and became "rebels" (zaofanpai) if their fathers had been classified artisans or upper middle peasants.[55] In Huatou, the factions grabbed arms and explosives from local depots, plundered shops, and fought a short battle, which destroyed two houses. Other young radicals were more timid: when rebels tried to topple Shi Dongzhu, the party secretary of Shiyan brigade, he sent them back to work: "Once you're finished with your work, you can come back and criticize me!" They obeyed: "Of course, they did," Dongzhu said, "I was a party member, and they were not. I supported Mao as much as they did; how could they take power from me?"

Industrialization and Deindustrialization in Jiajiang

By the end of the collective period, most people in the paper districts had become what they had never been before: subsistence farmers trying to wrest a living from impoverished and rapidly eroding soils. Paper output dropped from close to 5,000 tons in 1951 to just above 500 tons in 1977.[56] The industry survived, much reduced, in the parts of Hedong that continued to receive limited grain supplies, and in the more remote regions of Hexi, where state controls were so lax that some teams could engage in off-again on-again papermaking for the black market. Living standards in the paper districts dropped precipitously: in one example, both monetary wages and grain rations halved—from 1.54 to 0.81 *yuan* for an average workday and from a lavish 30 kg to the standard 14 kg per capita per month—after papermaking was phased out.[57] Perhaps the clearest evidence for the impact of rural deindustrialization comes from the demographic data: population growth rates of less than 10 percent in the years between 1951 and 1990, a time in which the nation's population doubled, are proof of long-term, grinding poverty.

Deindustrialization was, of course, only one part of the picture. "Walking on two legs," the simultaneous development of agriculture and rural industry, was a declared aim of the Maoist development strategy. Although the brigade and commune enterprises of the Maoist period pale in comparison with the spectacular rural industrial boom of the post-Mao period, they were held up as models for the developing world by foreign experts in the 1970s.[58] Jiajiang, too, saw rapid rural industrialization in the Maoist period, almost completely concentrated in the fertile plains around the county seat. Paradoxically, industrialization in the plains and deindustrialization in the hills were driven by the same logic: industrial development as resource mobilization in "small and complete" (*xiao er quan*) cellular units. The factor that connected the two processes—the "key link," in the Maoist phrase—was grain.[59] After the Great Leap famine, the central government severely limited the amount of grain that local governments could extract from farmers and redistribute. At a time when teams in the hills and mountains converted bamboo land to fields at enormous costs, reduced grain extraction enabled teams in the plains to take some of their best land out of grain cultivation and to plant tea, rapeseed, vegetables, and sugarcane instead.[60] On this basis, they branched out into food processing: the first commune and brigade enterprises in Jiajiang were flour mills, oil presses, and distilleries, all of them concentrated in the plains. Accumulated profits from these industries were then invested in brick kilns and other factories, again concentrated in the plains.[61] At the same time, reduced redistribution and the insistence on local autarky forced hill dwellers to abandon their specialization and become farmers, at great economic and human costs.

One of the declared aims of Maoist developmental strategy was to reduce spatial inequality, yet the effect in Jiajiang was the exact reverse. By insisting on a kind of formal equality—both hills and plains must become grain self-sufficient—the state exacerbated the existing inequality between the ecologically favored plains and the disadvantaged hills. This came on top of other measures that increased spatial inequality: the concentration of state industrial investment in the plains, mostly in townships just outside the county seat, and the building of roads and power grids in the plains long before the hills were connected.[62] Despecialization in the hills was irrational because it shut down an industry that made the most of an inhospitable environment, but did not replace it with anything remotely approaching its resource efficiency. Over the centuries, papermakers had developed efficient and sustainable ways to produce high-value industrial goods from land that was simply not suited for agriculture. Papermaking gave work to thousands of people and sustained larger populations than agriculture alone could have done. The opening of "wasteland" decreased rather than increased the number of people who could live off local resources: ten *mu* of bamboo land could keep a small workshop in year-round operation, but ten *mu* of agricultural land could not feed a family. Because agriculture could not sustain the same population densities as papermaking, full-scale agrarianization became an option only after the Great Leap famine had wiped out a quarter of the population. Even then, it came at the cost of dramatically reduced living standards. Ultimately, what lay behind the state's approach to papermaking was a failure of the imagination: a failure to consider the possibility that rural industry might be anything but a sideline or supplement—a "branch" (*mo*) to agriculture's "root" (*ben*).

7   The Return to Household Production

THE FIRST SIGNAL THAT THINGS were about to improve for Jiajiang's paper industry came in early 1978. Through an unpublished internal report of the Xinhua News Agency, Li Xiannian (then vice premier in charge of economic planning) learned about the dire problems papermakers faced. At a conference of the Light Industry Ministry in April 1978, Li criticized the county government and the ministry:

After reading the report, I felt that things really should not have been allowed to go this far. It is obvious that Lin Biao, and even more the Gang of Four, have severely disrupted and sabotaged [the industry]. However, we also have paid too little attention to the development of Chinese art-paper production. If this report is true, the production of Chinese art paper has not only not received active support—you may not like to hear it, but production has been attacked! Are there still individuals who wish to eliminate such a special product? The purchase price for Chinese art paper, for example, is ridiculous; [with such low prices] how can you raise enthusiasm for paper production? . . . I hope that our comrades from the Ministry of Light Industry, from Sichuan province, and from Anhui province take positive measures and earnestly implement our policies. Save Chinese art paper and make our national culture and art more flourishing! The people will thank you for this![1]

It is evident that Li was using the Jiajiang paper industry to settle scores with opponents of Deng Xiaoping's market reforms, of which he was a key supporter. Li's linking of handicraft paper to national culture and art and his hinting at leftist sabotage galvanized county officials. Within weeks, purchase and sales prices were raised, raw

158

material supplies were improved, and grain premiums were reintroduced. This last step was crucial, because higher cash returns alone
were not sufficient to increase production. Despite Li's pointed description of Jiajiang paper as "Chinese art paper" (*guohuazhi*), most
Jiajiang paper at that time was in fact coarse writing paper, known as
tuzhi (native paper) or *wenhuazhi* (culture paper). Officials in Jiajiang immediately began to adapt realities on the ground to Li's mis-
conception. Grain premiums for *guohua* paper were set at 100–150
kg of husked rice for 10,000 sheets; in contrast, premiums for *duifang*,
still the most common product, were set so low (10–15 kg per
10,000 sheets) that their effect was negligible. Moreover, the price
of *guohua* paper was increased by 26 percent, while the *duifang* price
remained unchanged.[2] The result was, as intended, a shift from *duifang* to *guohua* production, as teams in Hedong resumed or intensified production of large-format, high-end paper. Hexi producers
lacked the know-how to make *guohua* paper, and no guidance or instruction was provided to them.

When Li called for a revitalization of the paper industry, what
he had in mind was not a return to household production but a recentralization of the industry. In 1979, plans were drafted for a large-
scale factory combining mechanized pulp preparation with manual
molding and brushing. Despite generous funding, construction was
delayed until 1985, and when the factory opened, it could not compete with the newly privatized household workshops.[3] This and two
other attempts to build *guohua* factories led to the introduction of
know-how and machinery to Jiajiang, much aided by the fact that
some of the large paper mills elsewhere in the province had begun to
replace their equipment and sell used pressure steamers and pulp
beaters. This used equipment became the basis for a small-scale,
mechanized paper industry in Jiajiang, which grew in fits and starts
throughout the 1980s and 1990s. At the same time, a newly ener-
gized Light Industry Bureau began to push for technical changes
among rural papermakers. Soon thereafter, the new techniques
and machines were picked up by newly independent household
workshops, who put them to a much better use than the collectives. Household workshops quickly adopted pressure steamers and

diesel-powered pulp machines, which revolutionized the production process. Here, too, producers in Hedong were in a more advantageous position, since they received guidance from the Light Industry Bureau.

Handing Back the Workshops

Under the leadership of Zhao Ziyang, Sichuan was one of the first provinces to dismantle rural collectives and return land and other assets to rural households. Local governments were, however, often reluctant to implement the new policies. In Jiajiang, as in most neighboring counties, the collective structure remained intact until 1982, when the central government officially endorsed household production.[4] By the time collective land was distributed in Jiajiang, standard procedures were in place. All fields of a team were measured, graded according to soil quality, and divided into as many equal plots as the team had members, which were then distributed per capita, as "ration land" (*kouliangdi*). Land-rich teams in the plains distributed only part of their land on a per capita basis; the other part became "responsibility land" (*zerendi*), which was rented out through competitive bidding. Forest and bamboo land were divided in 1982–83, following the same procedures; in order to compensate for differences in the density and maturity of stands, households had to pay for the trees and bamboo they received with the land. The money from these sales was then equally distributed among all households.

The complexity of land distribution paled in comparison with the distribution of the paper workshops, which took place in 1983, a year after land distribution. Workshops represented the labor and capital investments of several generations, and frequent splits and mergers under the collectives made it all but impossible to find out who held a claim to a particular asset. In addition, decollectivization guidelines stipulated that workshops be kept intact and that equipment be given only to skilled producers. In Shiyan, the first step was to assign a price (usually below market value) to each piece of equipment. Next, workshops were broken up into vat units, each consisting of a vat with a press, a fixed number of drying walls, and other equipment. Small groups (*xiaozu*) were formed to match the number of these units, and lots were drawn to determine which group would get which vat. Any household not interested in papermaking could sell its share to other members of the small group at a price fixed earlier in the process. The one exception to this process was the distribution of drying walls, which represent the largest investment in the workshops. In contrast to buildings, ponds, and vats, many of which were inherited more or less intact from pre-collective workshops, all drying walls were of recent origin, since walls need to be rebuilt every two or three years. Like other equipment, drying walls should have been divided equally, but this would have led to a situation in which one family's drying walls would be in another family's home. Roaming teams of brushers that worked in other people's homes were common under the collectives, but this was considered incompatible with the new, household-based ownership structure. Instead, walls were sold at discounted prices to the household in which they were located, putting some households in a much better situation than others.

In the end, small groups of three or four households, usually neighbors or relatives, became joint owners of a single-vat workshop. Such groups were viable production units, similar to the first mutual-aid teams of the early 1950s. However, all of them dissolved within weeks, and all equipment became property of individual households. Such single-household workshops might consist of one-third of a vat, one-fifth of a steamer, half a paper press, and a few dozen drying walls. Contrary to what one might expect, there was no great enthusiasm for decollectivization. Since assets were distributed per capita, large families were usually in favor, but most people in Shiyan did not know what decollectivization would bring. The advantages of independent household agriculture were fairly obvious, but the success of paper workshops depended less on ownership than on access to markets and on the relative prices of raw materials and paper. At a time when markets were still in their infancy and most raw materials controlled by the state, independent production in small household units was not necessarily seen as the best option. In the end, privatization was pushed through by local cadres, amid much fear and trepidation. Cadres were as ambivalent about privatization as other villagers, but they felt that without a collective power structure to back them, their position had become untenable. As one village leader put it, "Our work became just too difficult. We did not care whether people agreed or not; we just wanted to return everything to the households and be done with it."[5]

Workshop Consolidation

Due to several unforeseen factors—liberalization of the grain and paper markets, increased urban demand for calligraphy paper, and the collapse of competing handicraft paper industries—privatized papermaking became a resounding success. Faced with growing demand and fragmented ownership of equipment, household workshops revived traditional methods of mutual aid and labor exchange. Distrustful of bureaucratically organized cooperation, people left their assigned small groups and turned instead to close friends. In one instance, a family that had received a vat but very little drying space joined forces with a family in a different team that lacked a vat but had sufficient drying space. This involved much carrying of pulp and paper over a distance of several kilometers, but both families felt that cooperation worked only if based on strong personal ties. One or two years after decollectivization, even these forms of cooperation had become rare. Some households had opted out of papermaking and sold their equipment; others had consolidated their workshops and become independent producers. Threshold costs were relatively low: vats and soaking basins, made from local sandstone, cost about 150 *yuan* each; paper presses 200 *yuan*; sieves, frames, and other small equipment about 100 *yuan*. The most expensive items were the drying walls, but these could be added gradually. Because of limited spending opportunities in the collective years, many households had considerable savings; those without savings could borrow from others who had liquid funds. Here, too, pre-1949 practices were revived: households could easily arrange short-term loans from friends or relatives, usually without paying interest. In the first years after decollectivization, rural credit cooperatives (RCCs) were eager to lend to papermakers. The Macun credit cooperative, for example, provided hundreds of loans in the range of 1,000 to 3,000 *yuan* to papermakers, at interest rates of around

8 percent a year, which was below the inflation rate. According to the rules, borrowers needed collateral and guarantors, but at a time when RCCs were flush with funds and there were few alternative investment opportunities in Jiajiang, these rules were interpreted quite loosely.[6] Credit for papermakers dried up after 1985, when a wave of township and village enterprise (TVE, *xiangzhen qiye*) construction absorbed all available funds.

The rapid revival of household-based papermaking was possible only because small household workshops swapped and pooled their resources in ways that allowed them to make the most of their limited means. This is most evident in the way households drew on one another in order to replenish their incomplete set of skills. Just as households inherited fragments of workshops from the collectives, they emerged from the collective period with fragmented complements of skills. With the decline of papermaking in the late collective years, opportunities for training had become rare. Apprenticeships in the workshops were assigned by team leaders, who tried to distribute jobs evenly between households. As a result, most families had one, but rarely more than one, member in the paper workshops. Novices were increasingly trained by master workers in collective workshops, who were unenthusiastic about teaching those who were not members of their families. Training was time-consuming, and under a piece-rate system, time devoted to instruction meant reduced income. Over the years, it became more and more unusual for a family to have a full complement of skills. After the dismantling of the collective workshops, people swapped and traded skills, just as they swapped vats and drying walls. Households with complementary skills often formed partnerships that were dissolved as soon as the partners had learned each other's skills.[7]

A Small Technological Revolution

The return to household production coincided with the introduction of time- and labor-saving technologies, many of which had been developed in the experimental workshops of the early collective period, almost thirty years earlier. Although falling short of full-scale mechanization, these new technologies profoundly transformed paper production in Hedong. They never reached Hexi, where workshops continued to make *duifang* paper with vats and steamers inherited from collective workshops. The first important innovation that reached Hedong was the pressure steamer (*zhengguo*): a steel container that could hold one to two cubic meters of bamboo or other fibrous matter, strengthened by a surrounding layer of reinforced concrete. The high temperatures and pressures achievable in these steamers, together with aggressive chemicals, reduce the steaming time from two weeks to eight hours. More important, preparatory "retting" (the soaking of the bamboo until the soft "flesh" has decomposed and only the long fibers remain) is no longer necessary. Instead, the bamboo is crushed and cut into short pieces and then soaked in caustic soda for 24 hours. Pulp machines, also introduced in the early 1980s, replaced foot-driven pulping hammers. Hollander beaters were too expensive for most papermakers, but mass-produced fodder cutters, used in pig farms to chop up silage, were an adequate and cheap alternative (Fig. 7). A third labor-saving device was the car jack: placed in a steel frame or under a lintel, car jacks could be used to squeeze out excess water from piles of freshly molded paper faster and more thoroughly than the old wooden paper presses.

Although usually bought secondhand, the new equipment was more expensive than wooden steamers, hammer mills, and presses. In 1985, used pressure steamers (discarded by urban factories because of the risk of explosion) cost 5,000 *yuan*, secondhand fodder cutters cost around 1,000 *yuan*, car jacks around 300 *yuan*. Steamers were beyond the financial reach of most households, but as noted in Chapter 1, steamers can easily be shared by several households. New-style *zhengguo* steamers were much smaller than the traditional wooden *huangguo*, but because steaming time had been reduced from two weeks to a single day, one pressure steamer can easily supply twenty vats with *liaozi*. In Shiyan, seven steamers, dispersed in the hills, serve 140 vats. Steaming *liaozi* has become a specialized service industry: for a small sum (30 *yuan* for a steamer holding two cubic meters, in 1998), a workshop obtains the use of a steamer and a soaking pond for a day, as well as fuel and the labor of the steamer owner. The unskilled work of filling and emptying the steamer is left to the contracting workshop, but most steamer owners insist on firing the steamer themselves. Most steamers have no safety valves and may explode if overheated. Indeed, one of the first steamers installed in Shiyan did blow up, killing its owner and a passerby. The owner's income, after expenses for coal, is 15 *yuan*—the same as the day rate for unskilled labor. This is not much, but steaming is light work and does not require constant supervision.

Papermakers also began to experiment with new raw materials. Declining bamboo supplies during the collective period had forced collective workshops to use dried instead of fresh bamboo; this yields larger quantities of pulp but with shorter, less elastic fibers. To counteract the resulting brittleness, long fibers from hemp, cotton, or the bark of paper mulberry or common mulberry were added. At the time of decollectivization, Hedong papermakers found a material that was not only cheaper than bamboo but contains fibers that are as strong and supple as those of the very best *baijia* bamboo: *suocao*, also known as *longxucao* (*Eulaliopsis binata*, sabai grass), which grows wild on deforested hillsides in Sichuan and Shaanxi. *Suocao*, which like bamboo can be harvested annually and produces steady yields year after year, was a cheap and plentiful resource. Moreover, it made paper of excellent quality.

Another raw material adopted at the same time was waste paper made from wood pulp, usually in the form of paper cuttings (*bianzhi*) from urban print shops. Although Jiajiang papermakers hate to admit it, waste paper quickly became one of the main raw materials in their industry. The better grades of waste paper do in fact result in good paper, although they need to be mixed with long-fiber stock such as fresh bamboo or *suocao*.[8] However, most producers who use *bianzhi* use it excessively and indiscriminately, not because it is cheap (good *bianzhi* is about 20 percent more expensive than bamboo) but because of the speed and ease with which paper can be made from waste paper.

Jiajiang is exceptional among the five or six large papermaking districts in China in its eclectic use of raw materials. Other handicraft paper industries pride themselves on using only traditional raw materials: Jingxian in Anhui, home of the famous *xuan* paper, uses rice straw from a special type of long-stemmed rice and the bark of a species of elm, mixed in fixed proportions; Fuyang in Zhejiang uses bamboo; Qian'an in Hebei uses mulberry bark. In contrast, paper-makers in Hedong combine long and elastic fibers from *suocao*, fresh bamboo, hemp, cotton, or tree bark with the shorter, more absor-bent fibers from dried bamboo or waste paper. Experienced paper-makers can maintain the quality of their paper while changing raw materials in response to fluctuating market prices. Few papermakers in Hedong still bother to grow bamboo; not only were old bamboo stands depleted in the collective period, but acid rain and a mysteri-ous disease have led to a rapid fall in output. Almost all raw materi-als used in Hedong now come from outside the production area: bamboo from Hexi and Emei, *suocao* from southwestern Sichuan and Shaanxi, waste paper from Chengdu and other cities, tree bark from Yunnan.

The main reason workshops responded so enthusiastically to the new technologies was that they greatly speeded the production cy-cle, allowing workshops to recoup their capital outlay in days or weeks rather than months. Traditional papermaking relies on the natural decomposition of the nonfibrous parts of the bamboo stalks, which takes about three months, and on natural fermentation of the *liaozi*, which takes several weeks. Collective workshops tried to com-plete three full production cycles in a year, but commonly achieved only two. Pressure steaming replaces these slow, natural processes with high pressures, high temperatures, and aggressive chemicals. The new steamers use about ten times more chemicals than the old wooden *huangguo* steamers, and they use caustic soda (sodium hy-droxide, $NaOH$) and sodium sulfate (Na_2S) instead of soda ash (Na_2CO_3), making papermaking much more polluting than it was in the past. Because of the reduced soaking time, papermakers can now sell the first batch of paper a few days after buying or harvesting bamboo or other fibrous materials. The small size of pressure steam-ers (0.3 to 0.5 tons, against the more than 10 tons of large collective *huangguo* steamers) also increases circulation speed. Large workshops now steam thirty small "pots" a year, instead of one to two large ones. Small workshops steam less frequently, but they speed up circulation even more by using waste paper, which can be pulped and turned into new paper within days. Quick turnover reduces papermakers' dependence on outside capital: in contrast to Republican-era paper-makers, who needed massive capital inputs at the beginning of the season, today's papermakers spread their sales and purchases over the entire year.

For cash-strapped papermakers, reducing the time between capital outlay and capital return is more important than saving labor. In fact, many papermakers claim that the new machines do little to improve overall productivity. This cannot be entirely true, since pulp machines release one laborer per workshop from the tedious operation of the pedal tilt-hammer mill, and pressure steamers do away with most of the labor in preparing *liaozi*. Based on interview data, I calculated a labor input of 0.8 to 1.2 workdays per *dao* (one *dao* is 100 sheets) of paper, depending on paper quality. This is sub-stantially less than the 2.3 workdays that it took to make 100 sheets of quality paper in the 1950s (see Chapter 1), with most gains due to reduced labor input in *liaozi* preparation and pulping. However, such gains are partly offset by the increased demand for managerial work in private workshops. Today, workshop owners spend much of their time purchasing raw materials and negotiating with traders, func-tions that in the past were carried out by collective leaders. Often, the labor freed by mechanization goes directly into marketing: household heads spend less time in pulp preparation (their tradi-tional task) but more in the marketplace.

Households, Government, and Technological Change

What explains the sudden outburst of technological creativ-ity in the 1980s, after a long period of stagnation under the collec-tives? Although the willingness of small household producers to ex-periment and take risks was the main driving force, it is important to acknowledge that local government action also played a role. Much of the initial impulse for change came, quite simply, from the ill-humored outburst of a powerful old man. But Li Xiannian's inter-vention would have had little impact without cumulative changes in transportation and infrastructure. The most important single change was the completion of the Chengdu–Kunming railroad, which passes through Jiajiang and links it to the national railroad grid. On a more local scale, villages were increasingly connected to power grids and transportation networks: by the early 1980s, most people in Hedong lived within an hour of a motor road, and by the mid-1980s, most Hedong villages had electric power. These were vital preconditions for the transformation of the paper industry. The new technological package of pressure steaming, strong chemicals, mixed raw materials, and pulp machines was originally developed for state-owned or mechanized collective workshops, but it was passed on to households free of charge. After Li's speech, officials of the county Second Light Industry Bureau (who opposed the dismantling of collectives, not only because it would put them out of work but also because they believed, correctly as it turned out, that abandoning collective controls would harm the reputation of the Jiajiang industry) worked hard to make the new technology available to household workshops in Hedong.[9] These workshops then shared their knowledge with other papermakers, and by 1985, traditional *huangguo* steaming had disappeared from Hedong.

Paper Production in Shiyan: A Sample

Since most of the following discussion of reformed household papermaking draws on fieldwork in Shiyan village, it may be useful to start with a short overview of the village and its economic structure. In 1995, Shiyan had 1,305 inhabitants living in 365 households. These were households as registered in the official household register (*hukoubu*). For internal use, village officials kept a separate count of "real" (income-sharing and co-resident) households, of which there were 310. The reason for this discrepancy is that stem families (families including parents and at least one married son) often register as separate households before they actually split up. According to official figures, average household size was 3.6, including 2.0 working adults; according to the unofficial count, household size was 4.2, with 2.4 working adults.[10] Forty-five percent of the income-sharing households (139 out of 310) owned paper workshops. Another 24 households bought and sold paper or raw materials, and 16 households derived their main income from wage work in the paper industry. Together, these three groups amount to

58 percent of all households. The village also included 21 factory workers, employed mainly in village-owned enterprises. In addition, there were 40 farmers, 27 artisans other than papermakers (mainly carpenters and bricklayers), 21 shopkeepers and small traders, eight village cadres, nine households without income (often elderly men or women who lived with their sons but were registered independently), and five households of unclear status.[11]

All paper workshops in Shiyan produced large-format (usually four-*chi*, 69 × 138 cm) calligraphy paper. Yearly output ranged from a few hundred *dao* to 2,500 *dao*, with most workshops in the 1,200–1,500 *dao* range. Self-reported workshop income ranged from 2,000 to 32,000 *yuan* a year, with incomes of fully specialized workshops clustered in the 15,000–20,000 range.[12] Despite my best attempts to check and recalculate the data, these are at best rough guesses. Only the largest workshops kept records of sales income, although most workshop owners remembered how much they paid for their raw materials and how much they got for their paper. In general, they were more likely to under- than to overestimate their income. Nonetheless, these were good incomes at a time when the average net income of rural households in Jiajiang was 3,148 *yuan*.[13]

The Reformed Household Workshop

In many ways, the production structure that emerged in Shiyan in the 1980s–1990s was similar to that common before land reform. Workshops were household-owned, and most work was performed by domestic labor, although all workshops supplemented their own labor force with hired labor. All workshops were also enmeshed in dense networks of cooperation and mutual aid, involving the exchange of labor, pooling of equipment, informal borrowing and lending, and similar practices. As in the past, steamers added a second level of organization to this household-based structure. Steamers were individually owned but served all workshops in a neighborhood and functioned as a crucial node in the circulation of information. However, these similarities obscure important changes. For one thing, papermaking in the 1980s and 1990s was much more strongly household-based than it had been in the past.

After decollectivization, all households started with roughly equal assets and equal chances. Almost all were undercapitalized, but most expected to build up fully functional workshops over time. In consequence, few men and women wanted to hire out as wage workers. On the employers' side, remaining fears that "the sky might change again" (that is, that the industry would be recollectivized and that employers of wage labor would be attacked as exploiters) discouraged workshop owners from expanding. In 1995, the village contained only sixteen wage molders or brushers, in contrast to the situation in the 1940s, when two-thirds of the village population hired out to a few big producers. Only 12 of the 139 paper workshops in the village employed full-time wage workers: one workshop employed three workers, two employed two workers each, and nine employed one worker each. The remaining 127 workshops relied on domestic labor, supplemented with hired short-term labor. All but one workshop in Shiyan operated a single vat; only one had two.[14]

Despite this seemingly flat distribution of equipment, workshops in Shiyan formed two tiers. Technical changes in the 1980s made it possible for young, labor-short, inexperienced households to engage in independent papermaking in ways that had not existed in the past. First, the use of pulp machines eliminated the need for a full-time pulper, reducing the minimum number of workers in the shop from three to two. Second, operations that used to take place in the workshop were outsourced to service providers: workshops no longer harvested and processed their own bamboo but bought dried raw materials delivered to the doorstep; the cutting and cleaning of raw materials could be left to hired workers; steaming had become a service industry; households without a pulp machine could have their *liaozi* pulped by neighbors for a fee. All this reduced the demands on domestic labor and made it possible for husband-and-wife teams to produce paper, albeit only intermittently. Most important, such small workshops could minimize their time and labor expenditure by buying waste paper, which was sold at an improvised paper market at Jiadangqiao. Waste paper could be processed in a few hours' time by soaking it in water with a handful of soda and some bleach, rinsing it a couple of times, and putting it through the pulp machine. Paper recycling allowed young families to engage in "hand-to-mouth" papermaking: whenever they had cash in hand, they bought a few bundles of waste paper, soaked and pulped it, and started molding. The first few batches of paper could be finished within days; with luck, capital outlays could be earned back in a week. Low-cost, quick-turnover production of *shengxuan*, as this type of paper is now called,[15] was a good way to start a workshop, learn the trade, and perhaps establish a few useful contacts. It was not, however, a long-term option. As producers gathered experience and accumulated capital, they shifted to "pure pulp" (*jingliao*) paper, which required higher skills and more labor input but had much higher profit margins.

Jiajiang was exceptional among Chinese paper districts in its total lack of standardization. Most of Jiajiang's competitors maintained some form of collective organization throughout the 1980s and early 1990s, which allowed them to standardize their products and maintain strict quality controls. The early and total transition to household production in Jiajiang made standardization impossible, with the result that Jiajiang produced a bewildering variety of paper, sold under fancy brand names such as "jade tablet," "dragon whiskers," or "imperial tribute paper." Many *jingliao* workshops produced for niche markets, specializing in unusual formats, extra-thick or extra-thin paper, highly absorbent calligraphy or nonabsorbent *gongbi* (meticulous brushwork painting) paper, paper sprinkled with mica or imitation gold flecks, or "imitation antique" (*fanggu*) paper, among others. Experienced producers can make almost any type of paper a customer wants and constantly adapt their products to changing market demands. Jiajiang producers also use their versatility with different raw materials and formats to imitate *xuan* paper from Anhui or to produce paper that looks and feels better at first sight than it actually is. This has led to a loss of consumer confidence in Jiajiang paper, not helped by the decision of some traders to market imitation *xuan* paper as *jiaxuan*—short for "*xuan* paper from Jiajiang," but sounding very much like "fake *xuan* paper."

For producers of quality paper, the optimum size of a one-vat workshop is still the same as it was in the past: six or seven workers. As in the past, the road toward high-quality, high-profit production leads through household expansion. In a sample of sixteen workshops on which I have detailed data, all but one producer of "pure pulp" *jingliao* paper employed at least three family members in their workshops; the average was 3.4. This is significantly more than in workshops producing *shengxuan*, which employ 2.8 family members. In another parallel with the past, adding household labor to the workforce makes it more, rather than less, likely that a workshop will hire long-term workers. Husband-and-wife workshops may need additional labor more urgently than larger workshops but generally cannot afford it. By contrast, a workshop with two working generations (usually the owner and his wife, and a son with his wife) may hire an additional vatman. As in the past, the added hired worker allows workshop owners to adjust the internal division of labor of the household: sons shift to high-skill, low-drudgery jobs such as pulp preparation, and the owner himself does little manual work apart from mixing the pulp and spends most of his time on marketing and quality control.

Riding the wave of household expansion has become more difficult in the post-Mao years. Since the 1970s, China has undergone "the sharpest fertility decline [ever measured] in any sizeable population."[16] At the same time, family division (*fen jia*), which used to take place only after the death of the senior generation, now tends to occur serially, as most sons insist on setting up their own households soon after they get married and have their first child. In the thirteen years from 1982 to 1995, household size in Jiajiang dropped from 4.46 to 3.45.[17] As mentioned above, households in Shiyan are larger than the county average (with an average size slightly above or below four, depending on the count used), and the number of adult workers per household is around two. But even this size is too small to meet the labor requirements of paper workshops. A logical response in such a situation would be to delay household division; after all, all household members would be better off materially in a well-staffed, efficient workshop. This is widely acknowledged, not only by senior papermakers, who would benefit most from keeping the household together, but also by the younger generation. At the same time, both young and old acknowledge that early division has become a cultural norm that overrides economic considerations.

This conflict is particularly acute in *jingliao* workshops, almost all of which include two generations of the same family. Almost all owners of *jingliao* workshops I interviewed complained about the laziness and extravagance of their sons, their lack of discipline, and their gambling. At the same time, they realized that they could not push their sons too hard without provoking demands for workshop division. Villagers expressed admiration for the few families that managed to maintain multigenerational workshops over many years, but generally accepted early division as inevitable. Village officials, who were often asked to mediate in family conflicts, strongly favored delayed division but also advised parents not to stand in the way of their sons.[18] Sons, on the other hand, echoed the post-Maoist critique of collective production in their discussions of undivided household workshops: in large workshops, they said, everybody "eats from the big pot" (*chi daguo fan*); brothers are treated the same regardless of their individual effort, and hard work is not rewarded.[19] Small units, by contrast, were praised as "giving full play to the enthusiasm" of owners who work for themselves rather than for a family collective.[20] The main source of conflicts in multigenerational workshops was not the work itself, which followed impersonal routines, but consumption-related issues. Fathers often complained that their son insisted on a motorbike or a TV set for his own nuclear family. Such purchases are seen as normal and necessary by the younger generation; senior workshop owners often see investment in the workshop as more pressing.

In most households, the customary "division by contract" has given way to a long-drawn-out process, in which sons split off successively and property is divided step by step.[21] While a household goes through this process, its boundaries are blurred, a situation described as "unclear division" (*fen jia bu qing*).[22] This opens a large field for conflicts. In one case I witnessed, the owner, a strong-willed and irascible man, decided in King Lear–like fashion to divide his workshop when he was only 45, so that he could "unburdened crawl toward death." Unable to live up to his declared intentions, he frequently clashed with his sons, who depended on him for paper sales (they sold their paper under his name) and for paper walls and other equipment (much of which he retained). At the other end of the spectrum, some families divide late or cooperate closely even after property division and reap material benefits from their cooperation. Take, for example, the four-generation, nine-person family of Shi Dingwen, who lives with his sons and their families in a large, undivided courtyard. For registration purposes, the nine persons in the family formed three separate households: Dingwen (then in his sixties) with his wife and his aged father, first son Yunquan with his wife and child, and second son Yunhua, also with a wife and child. Dingwen, his wife, and his father had their own source of income in a pressure steamer operated by Dingwen. His sons, both of them successful papermakers, have separate workshops, kitchens, and family budgets. For most purposes, the family formed two separate households: the brothers and their families ate separately (although visiting was common), and the three senior members ate with the sons in turn. At yet another level, the family functioned as a single household in which equipment was shared and members helped one another. Villagers hold up this family, and the two other four-generation families in Shiyan, as models to emulate. They are often described as "filial" (xiaoshun), a term that originally meant filial subordination but is now used to describe a joint effort by seniors and juniors to maintain intergenerational harmony.

Another consequence of China's demographic transition is that more families have only daughters, making it increasingly likely that workshops will be owned and headed by women. In the past, families without sons often adopted a close cousin, but this is no longer widely practiced. Of the five female workshop owners in the village, one was a widow whose husband had died when their pressure steamer exploded; two others were only daughters whose husbands had moved in with them. In such uxorilocal couples, both husband and wife insist that they own and manage the workshop together, but other villagers see the wife, not the husband, as the workshop head and use her name to refer to the workshop. Even more tellingly, in such workshops it is generally the wife that mixes the pulp and thus controls the paper quality. In one rather unusual case, a woman became workshop manager and household head (these functions typically go together) not by inheritance but by dint of hard work and talent. Che Suzhi was the leader of a women's team (funüdui) during the collective period and taught herself how to make pulp. When she and her husband opened a workshop in the 1980s, he became the workshop's vatman, and she mixed the pulp, supervised their wage workers, and took care of paper sales, with great success. Although only five of 139 workshops in the village are headed by women, two of these rank among the six or seven most profitable workshops in Shiyan.

Wage Labor

As mentioned above, long-term wage labor was used far less in the post-Mao years than it was before land reform. In 1995, sixteen households in Shiyan derived most of their income from wage labor. Since in some of these households both wife and husband hired out, the number of individual wage workers was around twenty. All of them were employed in Shiyan or neighboring Shijiao. Vatmen earned 2.0–2.5 *yuan* per *dao* of *guohua* paper, brushers 1.0–1.5 *yuan* for the same amount. Depending on the workers' strength and speed, monthly wages were in the range of 240 to 375 *yuan* for vatmen, and 120 to 225 *yuan* for brushers. In addition, workers received full board. Workers ate meat at least once a week (some employers served meat every other day); in addition, men were given as much cheap liquor (*baijiu*) and cigarettes as they could consume. If both husband and wife hired out, a couple could earn as much as a small-scale *shengxuan* workshop or more. In addition, their income was secure: demand for skilled wage workers was high, and workers who quit easily found work elsewhere. With high cash incomes and most of their daily expenses paid by employers, wage workers often have more disposable money than small household producers.

Nonetheless, wage work is not considered an attractive choice; in fact, it is often not a choice but the result of external pressures. Three out of the six wage-working couples I interviewed had a second child (in an area where the one-child policy is strictly enforced, and few young couples had more than one child) and had been ordered to pay fines. One couple had paid half of its fine (of 3,100 *yuan*) in the course of six years; another had paid no more than 30 *yuan* of its 3,420-*yuan* fine after four years. Nonpayment is possible only if a family has no assets that can be confiscated, and indeed, none of them had any property apart from the most essential household goods, and none reported any savings. Visible wealth in the form of equipment and savings in a bank account would almost certainly have been confiscated. Wage work, by contrast, put disposable income into the hands of these couples without exposing them to the risk of confiscation. As long as they spent their earnings, they could delay payment until the fines were canceled, forgotten, or eroded by inflation.

Labor Exchange and Mutual Aid

Mechanization and the introduction of service industries reduced the use not only of wage labor but also of labor exchange. Formal labor exchange as it existed in traditional *huangguo* steaming is no longer practiced, but informal exchange remains common. In contrast to market transactions between households (such as the hiring of wage labor and the purchase of services for steaming, transportation, or other tasks), which were readily discussed, unpaid exchanges between households were usually played down. Workshop owners routinely stated that they neither received unpaid help from others nor gave such help to neighbors. Yet this was plainly contradicted by the facts: in some cases, people stated that all labor in their workshop came from their own household, even as women who were obviously not members of the household sat in the courtyard, quietly picking through a pile of *suocao* or preparing paper for brushing. These helpers, we were told, "just happened to be around" and were making themselves useful, as female (but not male) visitors were expected to do. Many female tasks (*suocao* cleaning is a case in point) are time-consuming but require neither strength nor concentration; work of this type can be so casual and informal that it is hardly perceived as such.

The same claim of household self-sufficiency and independence was made about the use of drying walls. Because demand fluctuates so much, it is impossible for household workshops to match their equipment to production needs. Most workshops alternate between busy periods in which they are short of drying walls, and slow periods in which their walls are underutilized. Workshops dealt with this by borrowing and lending wall space from their neighbors according to their changing needs. Such exchanges were ruled by customary norms: no rent was charged, because it was assumed that the imbalance would in the long run balance out; recipients who felt that they received more than their due could express their gratitude by giving a cash present at the end of the year.[23] The same norms of reciprocity apply to cash loans. Loans of a few hundred *yuan* at a time were common between friends and relatives; they were usually paid back within months, and no interest was charged. Families that built houses sometimes borrowed several thousand *yuan* at a time, without interest (even at times of double-digit inflation rates!) and without collateral. Yet such loans, like other forms of mutual help, are mentioned only in response to very pointed questions.

The same insistence on the independence, self-sufficiency, and formal equality of small-producer households has been observed in other parts of China. Ellen Judd found that villagers in Shandong

strongly emphasized the separation of their respective households from one another. Members of households, and especially the senior women in them, were emphatic about the independence of their own households. It is unquestionably the ideal that each household should be self-sufficient, even while it is tacitly acknowledged that this ideal is impossible. . . . A keystone in this pattern is the strong expression that members of each household properly adhere to norms of being on good terms with *everyone*, but on especially good terms with *no particular person or household*.[24]

In Jiajiang, the norm of household independence coexists, in an uneasy tension, with often unacknowledged norms and practices that tie households to one another. On the whole, this combination has served Jiajiang papermakers well. The remarkably quick recovery of the industry in the 1980s was made possible by such low-key but nonetheless crucial exchanges between households. Household workshops started out with fragmented equipment and isolated skills but swapped vats, walls, tools, and know-how until most families had at least a minimally functioning workshop. Those with privileged access to new technology did not attempt to monopolize it but passed it on to neighbors. Technical information spread with amazing speed in the paper districts: new paper types and formats, new raw materials and additives, were introduced by paper traders or innovative papermakers and, if successful, were soon copied by other workshops. Cooperation and information-sharing remained rooted in kinship norms and practices that emphasized solidarity among agnates and distinction between generations.

Yet with the growing success of the industry, the boundaries of household workshops began to harden. Labor exchanges between households became feminized and therefore less conspicuous: the work of women who pick through a pile of their neighbor's *suocao* could be construed as play, and commitments that might otherwise arise out of such interhousehold flows of labor could be denied by workshop owners. Intergenerational harmony in extended household workshops was admired, but most people feel that such arrangements would not work for them. It was commonly assumed that the only way for sons to get along with fathers, and for brothers to get along with brothers, was to divide all common property; only people of exceptional strength of character could cooperate successfully with close relatives. The dominant image of the workshop was that of a self-contained, independent unit, in which all assets (including technological know-how and household labor) were controlled by an owner and his wife or, more rarely, by an owner and her husband. The limitations of this model are obvious: for technical reasons, handicraft papermaking requires large families, the routine use of wage labor, or institutionalized forms of mutual aid—or, ideally, a combination of all three. The ideal of small-household independence locked people into workshops that were too small to produce good paper and survived only through intense self-exploitation. At the same time, it made people shun dependent work, despite the facts that the income of wage workers compared favorably to that of small producers and that workers enjoyed greater income security.

In the light of changes since the late 1990s, it is tempting to link this ideology of household individualism to neoliberal discourses promoted by the World Bank and other international agencies, or perhaps more generally to the integration of China into a capitalist world economy. However, none of this was much in evidence in rural Sichuan in the 1990s. I would argue, instead, that these changes grew out of CCP policies from the 1970s on that regulated and to some extent *created* the rural household as an economic actor. As Judd has shown, the state not only regulated household residence and composition through its *hukou* registration system and thus created units that were more sharply bounded than the informal domestic units with which registration households overlapped, but it also assigned, through the household responsibility system, specific production tasks to individual households and regulated the economic rights and duties of "specialized households" and other state-recognized categories.[25] State regulation of the household as a unit of production was accompanied by a heavy dose of ideological remolding, aimed at promoting work discipline, thrift, civility, and, from the 1980s on, "market consciousness."[26] In Jiajiang, the impact of state "household-making" is evident in the broad acceptance of small households as the only viable units of production, despite the obvious economic drawbacks in the specific circumstances of Jiajiang, and in the condemnation of multigenerational workshops as products of an economically irrational collectivism. Ironically, the small, risk-taking, entrepreneurial household, perceived as natural and necessary, is economically irrational, as papermakers know only too well.

8   Paper Trade and Village Industries
in the Reform Era

THE CHANGES DESCRIBED in the previous chapter would
have been impossible without the spectacular expansion of paper
sales that took place in the 1980s and 1990s. Demand for paper increased rapidly in the 1980s, as the liberalized printing industry
turned out greater numbers of books and newspapers, enlarged state
bureaucracies and company headquarters used growing amounts of office paper, and the new urban middle class spent more money on
books and other paper products. Whereas the Maoist state had
aimed to fill the leisure time of citizens with political campaigns,
banning leisure activities from calligraphy to goldfish raising as
"bourgeois," the reform leadership encouraged people to pursue such
hobbies as painting and calligraphy.[1] Calligraphy had never been
subject to the same kind of attacks as other traditional art forms,
presumably because Mao and several other party leaders were ac-
complished calligraphers, but it was not widely practiced. In the
1980s, calligraphy regained its former function as a marker of moral
and intellectual accomplishment and of elevated status.[2] A growing
number of middle and high schools taught calligraphy, and millions
of children and adults started taking lessons in academies or from
private teachers. At the same time, the revival of popular religion
led to a growing demand for sacrificial paper. The 1980s and 1990s
saw a revival of large, public ceremonies in temples and lineage halls,
usually accompanied by the burning of huge quantities of paper. Religious ceremonies and private celebrations are also often accompanied by firecrackers, which created another growth market for hand-

made paper. Handmade paper delivers, quite literally, more bang for the buck, since the greater tensile strength of handmade paper results in louder explosions.

For much of the 1980s and 1990s, Jiajiang papermakers outperformed their competitors in Jingxian (Anhui), Fuyang (Zhejiang), Duyun (Guangxi), and Qian'an (Hebei), largely due to their better distribution networks. By the mid-1990s, Jiajiang natives had opened shops in all major cities of the PRC, and traveling sales agents supplied schools, academies, and stores in much of the rest of the country. A source from Jiajiang estimated that the county produced 40 percent of all handmade calligraphy paper in China, but none of the production districts kept reliable output statistics.[3] Whatever the precise number, Jiajiang in the 1980s and 1990s was clearly one of the biggest, if not the biggest, producer of handmade paper in the country.

Guerrilla Trading in the Early 1980s

Despite the suppression of the private paper trade in the collective years, the trade was never completely eradicated. Paper became a state monopoly in 1953, but private paper sales continued, now on the black market. After the Great Leap Forward and the breakdown of collective order, much of the paper trade slipped out of state control. Paper trading was brought back under control in the mid-1960s, but began to break away again during the Cultural Revolution.[4] Papermakers complained to their administrative patrons in the Light Industry administration about the "knife-and-whip policies" (*daobian zhengce*) of state trading firms: when demand was high, they were driven to increase output; when there was a glut, purchases stopped.[5] Yet neither knife nor whip could establish complete state control over paper sales. In Hexi, the majority of papermaking teams secretly contracted production to individual households, and even collective workshops sold part of their output behind the back of the state monopolies. Private traders (*zhi fanzi*) paid 30 percent more than state companies. Whether team leaders took advantage of these higher prices was a question of courage, not principle: leaders "with guts" (*danzi dade*) sold on the black market, whereas "cowards" (*danzi xiaode*) sold to the state.

The private paper trade was legalized in 1983, but it retained some characteristics of black market trading for much of the 1980s. Most traders engaged in a form of "guerrilla trading," moving quickly and over long distances and leaving as few traces as possible. Private traders operated at the interstices of the formal economic system: they were unregistered, untaxed, and neither policed nor protected by the state. At the same time, market demand and profit rates were high, and bank loans easily available to those with the right contacts. Several of the traders I interviewed spent most of the year on the road, taking their paper samples from city to city until they had filled their order books and then returning for a few days to Jiajiang to ship their orders. Only in the 1990s did paper traders adopt more orthodox methods of doing business. Rather than discuss the modes of operation of paper traders in general, I present a few sample cases.

Shi Rongxuan

Shi Rongxuan, born in 1932, worked as team head in a collective paper workshop and later became a purchasing agent (*caigouyuan*) for a papermaking team. After decollectivization, he gave his first batch of paper to a trader who promised to sell it on commission but never paid him. Rather than risk being cheated again, he decided to sell his goods himself. For several years, he took his paper to Chengdu by bicycle, over a distance of 100 km on poor roads. If he was lucky, he sold his load of 20 *dao* (60 kg) in market towns along the road and could turn back before reaching Chengdu. If not, he arrived in Chengdu on the evening of the second day, and sold his goods to passersby on Renmin nanlu, Chengdu's main avenue and the site of a nightly market for art, calligraphy, and antiques. If unsuccessful in Chengdu, he traveled on as far as Deyang city, 250 km north of Jiajiang. On his way back, he would buy 80 kg of paper scraps from printshops, which he sold to paper workshops in Shiyan. Such direct marketing was exhausting but profitable: in the early 1980s, he made a profit of 12–22 *yuan* on every *dao* he sold. After four years of itinerant trading, Rongxuan stowed away his bicycle and began selling by mail order. He kept his business small, selling to four or five retailers in Chengdu, who ordered by mail and sent him a 50 percent advance by postal remittance. Rongxuan then shipped his goods by bus, truck, or mail and received the remaining 50 percent of the sales price after his partners had received the goods. He estimated that he sold no more than 1,000 *dao* each year.

Shi Shengxin

Shi Shengxin was born in 1919 and worked as a vatman for Shi Ziqing. He had no children, and the son he and his wife adopted in the 1950s left him in the 1970s. When the collectives were dismantled, Shengxin and his wife found themselves as poor as before land reform, exhausted from a lifetime of hard work, and without children to take care of them in their old age. For a few years, they sold homemade bean curd for a meager profit. In early 1983, Shengxin put on his formal Mao suit and went to the Industrial and Commercial Bank in Jiajiang to ask for a loan. Being poor and almost illiterate, he was an unlikely candidate for a loan from a state bank, but he had heard through a friend that the ICB was looking for short-term investment opportunities for its unused capital. Shengxin proposed a partnership (*hehuo*) in which he would buy caustic soda from chemical factories and sell it to paper producers in Jiajiang. Amazingly, the bank offered him "several ten thousand *yuan*," for a 40 percent share of his profits. Since he could offer no guarantees, the bank dispatched a clerk to accompany him on his travels, pay his expenses, and keep track of his sales. The deal was illegal: banks were not allowed to give loans to unregistered private businesses, and the soda trade in Jiajiang was still controlled by the state-owned Materials Bureau (Wuziju). This, in fact, was the reason why Shengxin's proposal was so attractive to the ICB: as a monopoly, the Materials Bureau sold soda at such high prices that Shengxin could undercut them and still earn a huge profit for himself and the bank. Shengxin estimated that he earned 10,000 *yuan* in 1983, but in 1984 soda prices dropped, and Shengxin shifted to the paper trade. Like Rongxuan, he kept his business small, selling only to state or collective units (*danwei*) in the culture and education sector. Such units, he said, were the most reliable business partners; private entrepreneurs (*getihu*) were too likely to go bankrupt or simply pack up and decamp. For several years, Shengxin "ran the frontline" (*pao waixian*), that is, he traveled from town to town, selling paper. In the early 1990s, his adopted grandson convinced Shengxin to open a paper shop in Guangzhou, together with a small workshop for mounting (*biao*) paintings on scrolls. In the 1990s, Shengxin bought around 4,000 *dao* of paper every year, mainly from neighbors and friends, which he sent to his grandson. The shop also sold paper from Anhui, Guizhou, and other production districts, as well as brushes, ink, ink stones, and other paraphernalia for artists.

Shi Weifang

At the time of my fieldwork, Shi Weifang, born in 1951, was the most successful paper trader in Shiyan—as well as the proud owner of the only private car in the village, a Volkswagen Santana. Because of his bad family background (his father had been a *bao* head under the Guomindang), Weifang quit school early and worked for a time as a porter and a traveling beekeeper. Later, Weifang—like other men of his background—became a tailor, one of the few uncollectivized professions and therefore shunned by persons of good class background. After two years as an apprentice and ten years as a journeyman tailor, Weifang began peddling paper in 1982, when this was still illegal. His first trips took him to Chengdu and Leshan, where he sold his goods to schools and academies. After each trip he paid his suppliers, bought more paper, and left again. In 1984, he took out his first loan from the Macun Rural Credit Cooperative, enough for a trip to Beijing, Tianjin, and other northern cities. Carrying only a small sample book, he collected orders from paper shops, department stores, artists' associations, and academies. After collecting orders for several months, he returned home and sent the paper.

Most of his customers paid months after he sent them the goods. In contrast to other traders, Weifang obtained another loan from the RCC, which allowed him to pay a first installment to his suppliers. After his first few trips, money started to come in more regularly; at the same time, the RCC raised his credit limit and he could pay off his suppliers. By the early 1990s, Weifang had enough savings to pay cash for goods, but like most traders, he often bought and sold on commission. In a way, Weifang and other traders have reversed the *yuhuo* practice of the Republican period: instead of advancing cash or raw materials to papermakers, they take paper on commission and pay for it only after they have been paid by their customers, which may take half a year or more. Nonetheless papermakers were happy to sell to Weifang, who paid higher prices than most competitors and was considered the most solvent and reliable trader in Shiyan. Like many other traders, Weifang abandoned the traveling trade for direct sales through urban outlets. In the early 1990s, he opened wholesale stores (*kufang*) in Beijing, Tianjin, and Lanzhou; several other outlets were contracted (*chengbao*) to relatives of his wife, who, in imitation of a model then current in state and collective enterprises, paid him a fixed sum for every *dao* of paper that they sold and kept the remaining profit. Weifang claimed that he sold about 10,000 *dao* a year; people in the village estimated that he sold 10,000 *dao* every *month* and earned 100,000 *yuan* or more each year. Visible property is often the best indication of a household's wealth in rural China. Shi Weifang was one of the first people in the village to build a three-story concrete house and the first to own a car. He also paid most of the costs for a motor road and bridge connecting his house to the public road and bought urban residence permits for two children.

Peng Chunbin

Born in 1942, Peng is the son of rich peasants from Fanggou, near Shiyan. Despite his bad class background, he was sent to a vocational middle school. After his graduation in 1959, he went to Shiyan to serve as brigade cashier (*chu'na*). He married into a Shi family, learned to mold, and became head of a papermaking team. In 1964, he applied for a cadre training course. Instead of lying about his class background, as many others did, he admitted it, and that put an end to his cadre career. He resigned from his posts as team head and cashier and (like Weifang) became a tailor. After all private business was banned in the "Four Clean-ups" campaign, he worked alternately as a stone mason and a vatman. In 1970, he volunteered to go to the Liangshan Mountains, largely because he saw no future for himself in Jiajiang. Liangshan was a wild borderland, inhabited largely by (non-Han Chinese) Yi people who had been

"pacified" by the PLA only in 1957. For two years, Chunbin worked
in an arms depot in Puge county; then he returned to Jiajiang to
work as a tailor in the county seat. As in his previous jobs, he had to
pay a share of his income to his team in Shiyan to compensate them
for the loss of his labor. One of the advantages of being a tailor,
Chunbin said, was that most of his customers were cadres (few others needed to dress formally or could afford the service of a tailor)
who had access to first-hand information. In 1979, one his customers told him that household production would soon be legalized.
Chunbin's first impulse was to invest in a paper workshop, but after
some consideration, he decided to produce molding screens instead.
His mother-in-law taught him and his wife how to weave screens
from bamboo strips and horse hair and coat them with tree lacquer.
When private papermaking took off, everybody needed screens.
Since most of the old screen makers had died and few had trained
apprentices, demand for Chunbin's screens was high.

In 1983, his career took another turn when a delegation from the
Shaanxi Light Industry Research Institute visited Shiyan to study
handicraft papermaking.[6] Most villagers were wary of these strangers,
whom they suspected of wanting to steal their technology. Chunbin,
who was used to city ways, invited them to his home and made
friends with them. Soon after, he was invited to work as vatman and
screen maker in an experimental workshop near Xi'an. At the same
time, he started selling paper on the side. In 1989, he rented a paper
shop near the Forest of Stelae (Beilin) Museum, a place that attracts
calligraphers from all parts of China. Xi'an at that time, Chunbin
said, had a larger market for handmade paper than even Beijing, and
his shop prospered. Chunbin, his wife, and his five children bought
urban residence permits, and most of his children found work in
state enterprises. He still owned two houses in Shiyan: his old home
in the hills, which he let to a friend, and a new house with a large
storage space, conveniently located near the Jiadangqiao bridge. The
latter remained shuttered for most of the year, but received a received a steady stream of visitors whenever Chunbin was at home.
Chunbin remained emotionally attached to Shiyan, but the center
of his life has moved to Xi'an.

From Itinerant to Settled Trade

Although many Jiajiang paper traders continued to "run the market" (*pao shichang*) in the 1990s, the majority shifted to a more settled mode of trade by opening stores in the cities. Not only was settled life preferable to the hard and dangerous life on the road, it also allowed traders to reap the profits from retail sales, which before they left to local people. Most important, by settling close to their customers, traders were better able to collect debts, something that had always been difficult for itinerant salesmen. Yang Dehua, director of the Macun credit cooperative, estimated that in the mid-1990s, Macun natives owned some 200 paper shops in all parts of China. A frequently heard statement was that Jiajiang paper traders were active in all Chinese provinces "apart from Tibet and Taiwan."[7] Most shops were very small, consisting of a storeroom (*kufang*) and a small sales room (*menshi*), usually filled to the roof with paper and other items. In addition to paper, most of these stores sold ink sticks, ink slabs, and brushes, the "four treasures of the study" (*wenfang sibao*). Stores were often located near colleges, art academies, or places that attract tourists. In addition to retail sales to students, art lovers, and tourists, most shops supplied institutional buyers, which often accounted for a large part of their sales. Shopkeepers and their families often cooked, ate, and slept in the storeroom, in a few square meters of empty space between the piles of paper.[8]

Most of the paper traders I interviewed in shops in Chengdu, Xi'an, or Beijing complained about the random fees and bribes they had to pay police officers, the tax authorities, and other bureaucrats, as well as about the arrogance with which they were treated by many urban people. Rural Sichuanese formed one of the largest migrant groups in 1990s China; they were so numerous that people in other provinces referred to them as the "Sichuan army" (*Sichuan jun*), or, even worse, as "Sichuan rats" (*Sichuan haozi*). Sichuanese were generally considered hardworking and clever, but also boorish (*tu*), hot-tempered, and sly. Urban people often commented on their incomprehensible dialect and their small stature, which is associated with rural origins and poverty.[9] Paper traders often sought protection by cultivating local patrons, especially retired party cadres, army officers, and government officials who often become avid calligraphers after they retire.[10]

Relations Between Traders and Producers

As in pre-1949 Jiajiang, a large number of small traders competed for the products of an even larger number of producers. Once again, big traders tended to offer the best prices and conditions and tended to buy from big producers. The four biggest traders in Shiyan (Shi Weifang, Peng Chunbin, Shi Qingcheng, and Shi Rongjun) bought mainly *jingliao* from well-established workshops. Relations between them and their suppliers were stable and often involved loans: paper traders sometimes advanced cash to producers; more commonly, producers gave their paper on commission to traders. All the papermakers I talked to agreed that sales involving credit were preferable to impersonal transactions at the lower end of the market. Small producers of low-quality *shengxuan* paper sold small batches of paper to small traders and insisted on cash payments because small traders were likely to default. Such sales were characterized by mutual distrust and haggling. Trust is important in the paper trade because it is so difficult to determine the quality of non-standardized handmade paper. Paper traders usually remove a few sheets from a bundle, hold them against the light, tear off a small piece at the margin to examine fiber length and tensile strength, or touch the sheets with a wet finger to test water absorption. Yet ultimately, the test of the paper is in the painting—or in tearing, crumpling, or soaking it, all of which means that it can no longer be sold. Traders who buy from producers they know and trust spend less time on quality control; they also reduce the risk of angering or losing customers by selling substandard products.

Most sales of *jingliao* paper involved delayed payment. A common arrangement was for producers to deliver a first batch of paper and receive payment when they delivered the next batch a few weeks later. The sums involved could be substantial: Shi Shenglin, a big *jingliao* producer, said that he typically sold batches worth 3,000 *yuan* and was paid three months later. In effect, he extended a loan in kind to his buyers, who otherwise would have to borrow money at a monthly interest rate of 2 percent (the rate for low-risk loans at the Macun credit cooperative). Each such transaction saved the buyer 180 *yuan* in interest payments and increased the amount of liquid funds in his hands—a matter of great importance in the economy of the 1990s, when even profitable businesses were chronically short of cash. Papermakers accepted delayed payments from big traders because they knew that the situation could be reversed if necessary. Although Shi Weifang and other big traders were in debt to most of their suppliers most of the time, individual producers could always ask for prompt payment or cash advances when they were short of funds.

Township and Village Industrialization

Although papermaking recovered in the 1980s–1990s, it lost its former central role in the economy of Jiajiang. This was due in part to the growing concentration of the industry in two townships in Hedong. The Hexi paper district never made the transition to calligraphy paper but continued to produce *duifang* paper. In the 1980s, demand for *duifang* increased, both domestically (due to the revival of popular religion) and internationally (due to growing demand in Southeast Asia, where Jiajiang *duifang* was used to manufacture ceremonial paper money). *Duifang* producers also found niche markets in the manufacture of firecrackers and in print industries, where *duifang* was used as blotting paper. However, the *duifang* price collapsed in 1996 due to increasing competition from cheap machine paper. By that time, increasing demand for a large variety of "mountain products"—tea, medicinal herbs, timber, honey, and the dried buds of daylilies—had created alternative sources of income in the Hexi mountains. Year-round papermaking survived in a few villages (Tangbian, discussed in Chapter 1, was one of them), but most producers had reverted to a combination of agriculture, sidelines, and occasional papermaking. In 1983, Hexi still had more paper vats than Hedong (1,100 against 700); in 1997, most *duifang* producers in Hexi had gone out of business, and the majority of year-round vats were concentrated in Macun and Zhongxing townships in Hedong.[11]

As papermaking became concentrated in fewer locations and shrank relative to other industries, parts of Jiajiang underwent rapid industrialization of a very different type. The central factor in the rapid transformation of rural China in the 1980s and 1990s was the rise of township and village enterprises. Throughout the 1980s and 1990s, the TVE sector grew much faster than the rest of the economy, bringing wealth and employment to millions of rural people. By 1995, such enterprises employed 128 million people, produced one-third of China's gross national product, and provided one-third of China's exports.[12] Entire regions were rapidly transformed from agrarian backwaters to urbanized industrial districts. However, TVE development in many parts of China, certainly in Sichuan, remained rooted in the logic of state-led resource mobilization, a logic that often led to investment duplication and the concentration of factories in a few sectors. TVEs were also wasteful of land, energy, and water and often highly polluting. Christine Wong has described how TVE industrialization was driven by the need of township and village governments to generate fiscal revenues: "Faced with intense budgetary pressure, the current revenue-sharing scheme, and a tax system that remained dependent on industry for the generation of over two-thirds of total revenues, local governments have little choice but to engage in industrial expansion." At the same time, tax signals "ensured that much of the investment was wasted in duplicated and socially irrational projects."[13] She concludes that fiscal regulations created a convergence between the interests of local governments and enterprises that "is good for growth, as shown by the willingness of local financial departments to support investment [but] is not conducive to efficiency."[14] All these problems are evident in the village enterprises of Shiyan.

Blind Development: Village Enterprises in Shiyan

After the dismantling of the collectives, Shiyan was in danger of becoming what officials called an "empty-shell village" (*kong-ke cun*), a village without revenues and therefore without a functioning administration and incapable of providing even the most elementary services for its inhabitants. In 1985, the six members of the village party cell, led by party secretary Shi Shengyi and village head Shi Longji, began to search for a "project" that would generate revenue for the village. Given the long history of papermaking in Shiyan and the rising demand for paper, it is not surprising that they decided to build a paper factory. Papermaking promised immediate and stable returns and needed little investment. Plans were drafted by Liu Yuanzhi, a paper trader who had only a primary school education but had taught himself enough engineering and chemistry to design a semi-mechanized factory, consisting of three molding vats and a centralized pulp preparation unit with a large pressure steamer.[15] To raise money, all members of the village party cell pledged their personal property as collateral. This would not normally have been enough to secure the 160,000-*yuan* loan that they needed, but the director of the Macun Rural Credit Cooperative was a Shiyan man and a friend of Shi Shengyi and Shi Longji. The Xinghuo (Spark) Paper Factory started producing in early 1987; in the next two years, ten more vats and a hot-air drying chamber were added. Because of the larger overhead, production costs remained higher than in small household-owned workshops, but the Xinghuo factory's strict quality controls ensured that its paper fetched higher prices than that of any household workshop.

In 1993, Xinghuo gave up paper production and switched to saltpeter. Manager Liu Yuanzhi maintained that the factory was sound and that the decision to abandon papermaking was forced on him by county tax authorities. The market of the Xinghuo factory—fine paper for calligraphers and artists—was potentially profitable, but in order to establish a reputation among generally conservative calligraphers, Xinghuo had to renounce high profits for the first few years. This cautious strategy, Liu said, was not to the liking of township and county leaders, who in the hectic years after Deng Xiaoping's "southern tour" expected rapid growth. A more fundamental problem was that Xinghuo was classified as a handicraft workshop processing agricultural products and was therefore exempt from many taxes. Changing production lines would put Xinghuo into a higher tax category and raise its product tax (*chanpinshui*) from 6 to 30 percent of sales income. If increased revenue extraction harmed the long-term prospects of the enterprise, this was of no concern to tax officials. Liu said that he was accused of tax evasion and harassed by surprise visits until he agreed to the change. He admitted that he had evaded taxes but said that he was not alone: all township or village-owned factories did so with full knowledge of the tax authorities, and Xinghuo was singled out because of its product choice. Other people in the village disputed Liu's version of the events and said that Liu, too, supported the saltpeter project and changed his mind only after it failed. The converted Xinghuo factory remained profitable for two years, but the high demand for saltpeter attracted large numbers of competitors, leading to a glut in the market. After 1996, Xinghuo was no longer able to pay its bills and closed, leaving unpaid debts of around one million *yuan*.

The initial success of the Xinghuo factory encouraged village leaders to open a second factory. Few people in Shiyan had technical skills outside papermaking, and none knew how to manage a large factory. Village leaders therefore scouted around for a package deal, complete with technical blueprints, permits, state loans, and ideally a close relationship with a state unit that would buy their products. Having heard that a neighboring village had nearly landed a project for a strawboard factory, Shi Longji sent a delegation to Beijing to "hook up" with investors and administrative sponsors. After fruitless negotiations in Beijing, the delegation traveled on to a strawboard factory in Liaoning province, which promised to provide them with used machinery and blueprints. After the return of the delegation, Longji obtained support from the provincial TVE office, which was headed by a Jiajiang man, and from the banks. This left only the Jiajiang county government to be convinced. In a show of determination meant to impress the county government, the village leaders ordered the construction of a motor road and the leveling of 30 *mu* of bamboo land for the factory site. However, the county remained skeptical and sent its own fact-finding mission to Liaoning. In late 1987, the mission returned and reported that Jiajiang neither produced enough straw nor had a large enough market for a factory of this size. Shi Longji realized that he would have to find a replacement project or accept personal responsibility for the waste of money, land, and labor. Once again, he pawned his property for a trip in search of a project, this time to Shanghai. A few hours before his departure, he received a phone call from a friend, telling him about plans to build an acetylene factory somewhere in Leshan district. Longji immediately canceled his trip to Shanghai and set off to Leshan and then to the Economic Planning Committee in Chengdu, literally racing (if he is to be believed) against a delegation from Emei county that also competed for the project. Thanks to his driver, Longji was the first to arrive in Chengdu, and thanks to his connections, he secured the contract.

The acetylene factory, Longji explained, was a "fixed point" project (*dingdian xiangmu*). This meant that it was included in the state plan and that a large share of its output was contracted to state-owned construction units. Such a project, Longji said, could not fail; even if it lost money every year, the authorities would bail it out. The factory was fully automated; machines were installed and serviced by outside technicians; no village expertise was needed. Shiyan provided a location for the factory, close to a newly built motor road, and a workforce of 37 shop floor workers and 23 administrative staff and salesmen. In 1989, the first year of the factory, demand was low because of the harsh austerity measures imposed by the central government.[16] In the next few years, however, a building boom brought high profits to the factory, until once again overcrowding led to a glut and a collapse in market prices. In 1998, the factory had unpaid debts of more than two million *yuan*,[17] although by 2001, it was again making a profit.

In 1993, the village obtained the go-ahead for a third project, a calcium carbonate ($CaCO_3$) factory. Calcium carbonate, or chalk, is widely used in the construction industry. Usually obtained from limestone quarries, it can also be made from slaked lime, which is a by-product of acetylene production. The apparent success of the two existing factories convinced the banks and the county government that Shiyan's leaders had the skills to handle an even larger project. Fixed investment for the new project was four million *yuan*, more than twice that of the acetylene factory. The acetylene factory, itself still heavily in debt, stood security for the new loan.[18] During the first trial run in early 1995, it became apparent that the factory could not produce the high temperatures needed to produce calcium carbonate. The general opinion in the village was that "someone from our side has taken a commission" (*women zhefang de ren chi le huikou*)—in other words, that one of the village leaders had been bribed into accepting substandard equipment. The factory, which occupies about two acres of good agricultural land, has never produced anything and is now slowly decaying.

A fourth factory that was for some time owned by Shiyan lies outside the village, in neighboring Bishan. Since the 1970s, Bishan had produced roof tiles and bricks in small kilns, but it had no village-owned factory. In 1990, the Bishan village committee began drafting plans for a mechanized brick factory. The project was opposed by the Macun township government, which had just opened a brick factory of its own. A Bishan native who worked for the Leshan city government advised his village to "show courage" (*dadan de gan*) and increase the scale of the project. As he had predicted, the scaled-up plan won them the support of the county and Leshan district governments, who promised to arrange a loan. The village committee then began to raise funds from villagers and started construction, but the promised loan never materialized. In 1995, the village committee gave up all hope that it ever would and sold the half-completed factory, at a loss of more than 20,000 *yuan*, to a partnership consisting of Shiyan's acetylene factory and a private investor. Soon thereafter, the acetylene factory sold its share to the private investor. Bishan village cadres claimed that Shiyan, with its good contacts to the Macun RCC, ensured that Bishan's application for a loan was turned down and thus forced Bishan to sell its project. Once Bishan put the project on the market, the RCC channeled the money they had denied to Bishan to Shiyan, allowing Shiyan to snap up their project at a bargain basement price.

Gains and Losses from TVE Development

Throughout the 1980s and 1990s, village leaders in Shiyan devoted most of their time and energy to the expansion of village-owned enterprises. Their success during the 1980s and early 1990s earned them praise and material benefits and aroused the envy of neighboring villages. Most villagers, however, derived little benefit from the enterprises.[19] Data on enterprise profits are difficult to come by, but the physical evidence—two shuttered and decaying factories in the village, and one that operates irregularly—suggests mixed success at best. Village leaders estimated that the total debt of the three factories amounts to six million *yuan*. Over the years, the village factories contributed in a modest way to village expenses: they subsidized cadre salaries and underwrote relief measures for poor villagers, repairs of the village school, construction of a road and two bridges, and renovation of a historical building.[20] Together, these expenses amounted to about 200,000 *yuan*, spread out over several years—not insignificant in absolute terms, but less than 1 percent of the yearly industrial output (including papermaking) of the village.

One of the reasons, apart from the need to generate revenue, why local governments in the 1990s promoted TVE industrialization was that it provided employment for local people. At their peak, the village factories employed about 110 workers: 40–50 workers and 10 office staff at Xinghuo; 37 workers and 23 office staff at the acetylene factory. Only one-third of the factory workforce came from Shiyan; the rest came from neighboring villages and townships. In 1994, the peak year, the factories employed 7 percent of the village workforce, against 65 percent in the paper industry.[21] Work in the factories was not overly attractive for Shiyan villagers: although it was less back-breaking than work in the paper workshops, it paid less. Office jobs and jobs as sales agents, drivers, cooks, or security personnel were coveted, but there were only about thirty of those, and years could pass without a vacancy.[22] Another disadvantage of work in the factories was that wages were paid only when the factories were in funds. Xinghuo and the acetylene factory were often so short of circulating capital that they could not pay their electricity bills, let alone their employees.

Administrative and Economic Bifurcation

Technologically and socially, the village factories were transplants from the urban sector. Their physical appearance was modeled on state-owned work units: surrounded by high walls, they are accessible only through heavy iron gates, flanked left and right with boards that bear the name and administrative affiliation of the factory. Factory personnel, especially the clerical staff, were set apart from the village world by dress codes, work routines, and leisure habits. Workers wore white lab coats, similar to those used in urban shops and factories. They had a standard 40-hour work week, often much reduced by power cuts, with regular work hours and weekends off. Different terms were used for the work of factory employees and that of papermakers: papermakers and farmers "worked" (*gan huo*) or "hired out" (*da gong*), whereas factory workers "had a job" (*gongzuo*) or "went on duty" (*shang ban*).[23] Factory employees were the only people in the village who enjoyed structured leisure time. In contrast to hired molders and brushers, who took short breaks at irregular intervals and ate in haste and silence, factory workers enjoyed long and delicious meals in the factory canteen, followed by two-hour lunch breaks in which they took naps, played cards, or challenged one another to basketball or table tennis matches in the courtyard of the factory. Workers went on shift during the frequent power cuts, which sometimes continued for weeks, and spent their days chatting or playing cards until the early morning hours.[24] Generous leisure time and the pursuit of hobbies set the factory workers apart from the village population and gave them a distinctly urban flair. This was evident in a sports festival, organized by the township, in which teams from different townships competed in events such as table tennis, basketball, and running. Shiyan was represented only by factory workers, who alone had the leisure time to prepare for the competition.

The people most intensely involved with the factories were the village officials. Most officials in Shiyan had dual appointments in the administration and in the factories; it was thanks to factory employment that most of them were reasonably content with their work. Village and factory functions overlapped, but officials tended to see their work as factory managers, salesmen, or accountants as more meaningful and interesting than the routine administrative tasks in the village. In their dealings with the outside world, village leaders usually wore their business "hat," introducing themselves as factory directors rather than as party secretaries or village heads. Most village officials spent the majority of their time in the acetylene factory, far from the hills where most villagers live but close to the Jiajiang–Hongya road and well connected to the county seat. In 1996, the village committee opened a second office in the abandoned Xinghuo factory, somewhat closer to the center of village life. All business that required interaction with the outside world (fac-

tory business, meetings with township officials, and the like) was carried out in the acetylene factory, whereas such routine tasks as household registration, conflict mediation, tax collection, and birth control were handled in the Xinghuo office. The latter was staffed by only
two persons: the village head and the chairwoman of the women's
federation. Both complained about their work, which brought them
into frequent conflicts with co-villagers; the village head, in fact,
stopped attending to his duties and retained his position only because
nobody wanted to replace him in this thankless task.

Household Factories

TVE industrialization in Shiyan was almost the polar oppo-
site of household-based papermaking. Whereas papermaking was
rooted in historically grown structures and driven by market demand,
TVEs were transplants from the urban-industrial sector and driven by
a logic of state-led resource mobilization in which contacts with state
sponsors were far more important than market demand. In the mid-
1990s, a new kind of small-scale, household-owned, mechanized paper
factory emerged that seemed to bridge these poles. In Shiyan, the new
technology was pioneered by Shi Longji, the village head and factory
manager who had led the village search for factory "projects." In the
difficult first year of the factory, Longji was ousted in an internal
power struggle. A man of restless energy, he turned toward smaller
projects that he could run on his own or as partnerships with outside
investors. His first two ventures, a chicken farm and a small factory
producing electric boilers, failed and plunged him deeper into debt.
However, his paper workshop continued to produce some of the finest
paper in Jiajiang, which sold extremely well and allowed him to pay
off at least some of his debts. In 1995, his son, a technician in the
county paper factory, proposed building a small, mechanized paper
factory. Longji borrowed 50,000 *yuan* from friends and, with the help
of his son, built a continuous papermaking machine entirely from
scrap metal. It closely resembled the first paper machine, patented in
1806 by the Fourdrinier brothers of London. In Fourdrinier machines,
the stock (diluted pulp) is fed onto an endless belt; as the water drains
through the moving belt, the fibers start to mat. The wet paper then

goes through press rolls that squeeze out excess water and passes over a heated dryer roll, in this case a steel drum filled with burning coal. It is then wound on a roll and later cut to size. The end product is cheap *shengxuan*, sold at a price slightly below the cheapest handmade paper.[25]

Longji's example was soon followed by other papermakers in Shiyan. With a daily capacity of 1.5 tons, the next mechanized workshop to open in Shiyan was five times larger than Longji's and equipped with a factory-bought paper machine and steamer. The fact that it included a pressure steamer gave the owner greater flexibility in the use of raw materials: waste paper, the main input for this type of paper, can be mixed with stronger fibers made from *suocao* grass, straw, or bamboo.[26] Similar workshops have sprung up in other papermaking villages. They are unlikely to replace traditional manual workshops, since artists and calligraphers insist on handmade paper for their art work, although some may use machine-made paper for daily practice.[27] However, they do compete with cheap *shengxuan* paper and may in the long run force small *shengxuan* workshops to move up-market into the already crowded market for quality paper.

Proprietary Technology

Different from traditional paper workshops and similar to the village-owned factories, these new "household factories" are surrounded by walls and can be entered only with the permission of the owner. Shi Longji (admittedly atypical because of the tensions between him and other village leaders) built walls around his factory that were as imposing as those of the acetylene factory and did not allow visitors. He made an exception for my Chinese colleague and me, with the stipulation that we not take pictures or notes. Other owners of household factories were less defensive, but their premises were also walled and gated. In recent years, some of the larger handicraft producers also started to build walls around their workshops. One reason for this is that the construction of motor roads has made it possible to bring bricks to the hills, which have replaced wood as the cheapest and most common building material. Walled workshops provide protection from the elements and can be heated; they also provide protection against theft in workshops that now own more, and more expensive, equipment than in the past. Yet it is hardly a coincidence that the first workshops to build walls employed unusual and valuable technologies that might invite imitation by their neighbors. The workshop of Shi Xiujie (one of five female workshop owners in the village) produced six-*chi* paper (97 × 180 cm), molded by a two-man team; the other walled workshop had developed a new drying technology. Workshops using standard techniques, by contrast, remain open to view from all sides. Another illustration of the increasingly proprietary attitudes toward technology comes from Jinhua village, not far from Shiyan, where the party secretary discovered a new way to dye paper in brilliant colors. Instead of sharing the technology, he teamed up with the traders who supplied him with dyes and loans and bought his paper and built a network of subcontractors who buy his dyes, make paper according to his specifications, and pass it on through him to the merchant-investors. The process was not strictly speaking secret, but the control over raw materials and sales was concentrated in this man's hands and subcontracting households were told not to share their knowledge with outsiders.[28]

Changes Since 2000

In 2003, the county government ordered all steamers in the paper districts closed and restricted steaming to a single location just outside the county seat. This was a large walled area, divided into lots that were rented to about twenty operators who had lost their steamers in the hills. Papermakers now trucked their bamboo or *suocao* to the steamer site, where it was processed for a fee. Once it was finished, they picked up the *liaozi* in a rented truck or tractor and carted it to their workshops in the hills. Most workshops in Macun and Zhongxing had access to motor roads; they could continue to make paper, albeit at higher costs and reduced profit rates. Practically all papermakers in Hexi, and those papermakers in Hedong whose workshops were accessible only by mountain paths, were left with only two options. They could make paper without steaming, using waste paper as raw material, although this would condemn them to very low profit rates. Or they could lie low for a while and then secretly rebuild their steamers, evading this policy as they had so many others in the past. In 2003, however, all indications were that evasion would be difficult. Rumor had it that the county government had threatened to sack the party secretary and township head of any township that continued to harbor any unauthorized steamers, and smoke and emissions from the steamers make them impossible to conceal.

Behind this lay a drive to reduce pollution from the paper workshops. All paper production is polluting, but handicraft papermaking before the 1980s had relied largely on biological processes to break down the lignin and loosen the plant fibers. By contrast, the new pressure steamers with their aggressive chemicals produced much more concentrated effluents that made the mountain streams run black under a thick layer of cappuccino-colored foam. Water from the hills is used to irrigate rice fields in the plains, and since the 1980s, farmers had complained that steamer emissions were killing their plants. The county government ignored these complaints, but the impending completion of the Three Gorges Dam focused the attention of provincial and central governments on water quality along the Yangzi and its tributaries. Unless pollution was stopped, it was feared, the reservoir would turn into a cesspool. A ten-year plan to reduce pollution was passed, and all paper mills and other heavy polluters along the upper reaches of the Yangzi were ordered closed unless they met strict environmental standards. Since building water treatment plants for single workshops was not feasible, the county decided to centralize steaming—the most polluting process—at a single site.

From the point of view of the county government, centralization also had other advantages. Officials formerly involved with the paper industry had long argued that decollectivization had gone too far. Since the demise of the collectives, they argued, the industry was plagued by cut-throat competition, a loss of consumer confidence, and an unsustainable use of resources, problems that could be solved only by a partial recentralization of production and an imposition of administrative controls. Such proposals usually involved a three-pronged approach: the establishment of a research body that would conduct market and technological research, quality control through a state-licensed body that would certify paper that met its standards, and a trading company with a monopoly over certified paper. Officials also proposed the establishment of a centralized paper market in Jiajiang city, a unified trademark for all Jiajiang paper, and a paper research association.[29] Apart from a short-lived recentralization of the *duifang* trade in Hexi under one of the state trading companies, these efforts failed. Bureaucratic entrepreneurs could not outperform private competition, and no longer had the power to impose their will on household workshops. As the director of one of the trading companies put it, "people did not listen because we were no longer supported by the higher levels."[30]

After the disbandment of the collectives, the paper industry—potentially a source of considerable revenue for the county—was no longer systematically taxed. Paper workshops were classified as rural sidelines and as such were not taxed at all. Before the 1994 tax reforms, paper traders paid sales taxes; after that, they paid income and value-added taxes. In reality, the county lacked the means to keep track of paper sales and traders' incomes; it therefore negotiated a "quota tax" (*ding'eshui*) with individual traders. The result was that even as the paper industry expanded, tax revenues from the paper industry contracted from about 10 percent of the county's total income in 1991 to 6 percent in 1997.[31] By concentrating paper steaming in a single, easily supervised site, the county put itself into a position in which it could once again tax the industry.

These changes took place in the context of extremely rapid economic development in Jiajiang. Since the 1980s, the county government had single-mindedly pursued the aim of large-scale industrialization, in ways that bear more than a passing resemblance to the Maoist past. In three distinct waves, the county promoted brick kilns and factories for construction materials, large paper mills, and factories making ceramic tiles. During each wave, officials were told, among other things, to mobilize all available resources for these industries, to grant their requests for permits, help them find construction sites, and extend loans. Officially dubbed the "three great industries" of Jiajiang, they were mocked by old cadres in the Light Industry Bureau as the "three failures" (*sanci daomei*).[32] They also left the Jiajiang plains pockmarked with half-finished or deserted factories. Around 1999, however, the ceramic tile industry began to take off, aided by a construction boom in the Chinese interior and by the influx of capital and expertise from the ceramic tile industries of Foshan in Guangdong and Zibo in Shandong. At the same time, the tile factories—most of which had started as township enterprises and had remained collective owned during the difficult second half of the 1990s—were privatized. Ties between the private enterprises and the county government continued to be close, as the county assumed new roles as facilitator and provider of support for the tile factories and for potential investors. Policy documents from the first few years of the twenty-first century called for the establishment of a new spirit of "daring to do things in a big way" and of "being rich and wanting more" (*fu er yao jin*), and for the simultaneous destruction of the "negative peasant mentality" of being content with small steps and small wealth.[33] They also demanded the removal of bureaucratic hurdles standing in the way of investors and the transformation of all branches of government into service providers and facilitators. Investors were promised cheap loans and subsidies, as well as a safe investment climate in which "the government will do a good job in discovering and dissolving social contradictions, and disturbances that involve the masses will be nipped in the bud."[34] This policy paid off: Jiajiang now boasts several hundred ceramics factories and styles itself the "ceramics capital of western China" (*xibu cidu*). The red sandstone of the Jiajiang hills, which had in the past supported abundant bamboo growth, turns out to be an excellent material for tile production, and stone quarries are rapidly expanding into the hills.

9   The Jiadangqiao Stele

JIAJIANG PAPERMAKERS EMERGED from the collective pe-
riod relatively poor but also relatively equal. In 1983, all household
workshops started with little labor, little equipment, little experience, and few commercial connections. Over time, a two-tiered production structure evolved, with a large number of small producers in
the cheap *shengxuan* sector, and a much smaller number of large
workshops in the *jingliao* sector. The gap between the two was huge:
average profit rates for each *dao* of *jingliao* were twice as high as for
shengxuan, and the average *jingliao* workshop earned three times
more than the average *shengxuan* operation.[1] In addition, market
demand for *jingliao* was relatively steady, whereas *shengxuan* workshops had to deal with rapid fluctuations in demand. However, inequality was not perceived as permanent, since economic success depended to a large extent on family dynamics. Households that added
sons, daughters, and daughters-in-law and above all managed to prevent the workshop from splitting up could increase profits exponen-
tially. This became more difficult in the 1990s, when early household division became the almost universal norm. At the same time,
early division had a leveling effect, as large, multigenerational workshops became rare.

The logical consequence of decreasing household size would have
been an increasing use of wage labor, but this remained relatively
rare, partly because young couples preferred to start their own workshops, and partly because of a reluctance to hire more than one or
two workers per workshop. This reluctance was sometimes expressed
in political terms—the fear that "the sky could change again"—but
more than that, workshop owners worried about provoking the envy

of their neighbors. The implicit threshold, never passed in Shiyan during the 1990s, was parity between household and wage labor: a workshop employing roughly equal quantities of hired and domestic labor could still be seen as a family concern; once hired labor contributed more than 50 percent of the work, it became a capitalist enterprise. Since neither the domestic workforce nor the number of hired workers could readily be increased, workshop owners turned toward external service providers such as steamer operators or transport workers and increasingly toward mechanization—solutions that appeared more compatible with the dominant vision of independent and roughly equal small-producer households than the use of wage labor. Pulp machines were perfect in this regard: they were affordable (used pulp machines could be had for 1,000 yuan), and each machine replaced one semi-skilled worker, making it possible for a two-worker household to remain in business.

The same trend toward differentiation, and the same struggle to contain growing inequality, can be observed in the paper trade. Success as a trader was not a function of household size (although a large network of relatives outside the village had proved helpful to several traders) but of timing, since those who started in or before 1983 were most likely to succeed. For paper traders, there was no obvious limit beyond which expansion became problematic; rather, there was a complex trade-off between success and social obligations. As in the pre-1949 past, people in Shiyan stressed the central importance of xinyong (trust or credit) in the paper trade. Given the choice, most papermakers preferred to sell to the big traders known to "practice xinyong": men like Shi Weifang who had never actually defaulted on his payments (although he, like all other traders, often delayed payments for the paper he sold on commission) and who "looked after" (zhaogu) their suppliers by buying from them year after year. Once such a reputation was established, it became self-reinforcing, as more suppliers brought higher profits and increased solvency. However, creditworthiness had to be demonstrated by occasional cash advances and, above all, by contributions to community expenses: collections for villagers in need, construction of roads and bridges, renovation of the school building. And like big workshop owners, big paper traders were concerned that their wealth might invite theft, slander, or sabotage from envious villagers. Shi Weifang and other traders in Shiyan claimed that no such thing would ever happen to them, but they had heard of theft and violent attacks on rich people in neighboring villages. Crime, linked in most people's minds to outsiders and urban influences, was a growing concern.[2] Prostitution was much in evidence in the county seat, and just outside the village, heavily made-up women sat in front of a karaoke bar, waiting for customers along the Jiajiang–Hongya road. Many villagers in Shiyan linked crime and social tensions to the decline of kinship norms. In 1993, the Shis renewed their generation names and erected a stele inscribed with the new names in the center of the village. If honored by future generations, the new name list will ensure the existence of the Shis as a named, identifiable, and structured group until at least the late twenty-sixth century.

The Stele

In 1992, Shi Dingsheng's mother died. As customary, Dingsheng, a wealthy paper trader, party member, and secretary of the acetylene factory, invited his relatives and neighbors for the funeral. During the wake, Dingsheng's guests recalled the devotion of the deceased (who was not a Shi, of course) to the Shi family. From there, the discussion moved on to what many Shis perceived to be a decline of agnatic solidarity in the village. Families split early; sons refused to provide for their aging parents; sons and their wives verbally and sometimes physically abused their parents. These sentiments were perhaps strongest among those old enough to remember the annual caning of unfilial sons and daughters at the Qingming festival but were by no means limited to them. Most of those present, including the young and the middle-aged, believed that people needed to know who was senior and who was junior, and that this knowledge was slipping away. It was not that the list of twenty names fixed in the early Qing was running out: most Shis presently alive belong to generations eleven to fourteen, marked *ding, sheng, gui*, and *quan*; the very first *tais* (generation fifteen) had been born in the early 1990s.[3] Even if the Shis took no action, the first "nameless" generation would not be born until the middle of the twenty-second century. However, young couples were increasingly following the fashion of using single-character names for their offspring, or even naming them after Hong Kong or Taiwanese pop stars.

Dingsheng and his friends therefore decided to set up a committee that would choose twenty new generational names and inscribe them in stone, just as the three founding brothers had done. Apart from Dingsheng, the committee included other members of the senior *ding* generation: Shi Dingliang, the former land reform activist and commune party secretary, and Shi Dinggao, a former party secretary of Shiyan brigade. The driving force, however, was Shi Guizhong, a young and generationally junior paper trader. The committee's first move was to request permission from the Macun township government, which replied that it "neither supported nor opposed" the plan. This may not sound overly enthusiastic, but the same words (*bu zhichi, bu fandui*) had been used by Zhao Ziyang to express tacit support for the return to household farming and were understood as a clear go-ahead. Next, the committee began to collect donations. Virtually all the more than 500 Shi families in five adjacent villages—Shiyan, Shijiao, Bishan, Jinhua, and Zhangyan—contributed. The Qiliping Shis, a small group that had moved to neighboring Hongya county 150 years ago, also sent a donation, thus reinforcing their ties with the Macun Shis. The largest donations (100 *yuan* each) came from the party secretaries and village heads of Shiyan and Shijiao village and from several wealthy paper traders. Most other households contributed between 20 and 50 *yuan*.

Meanwhile, Shi Dingliang traveled to Mianzhupu, the original home of the Shis, some 40 km to the south of Jiajiang, to look for information on the Shis' history and, with the help of Dingsheng and Dinggao, compiled a new list of names and a short commemorative text. The committee then commissioned a stone stele, about three meters high and one meter in width and depth (see Fig. 12). Three of the faces were inscribed with the names of all donors; the fourth contained the two lists of generation names (the new list of twenty names begins with the characters *ke* and *xue*, which together form the word for "science") and Dingliang's text.[4] The first section of the text (translated in Chapter 2) discusses how the first ancestor and his wife moved from Xiaogan township in Hubei to Sichuan, how the first ancestor died before reaching the Sichuan basin, and how his widow gave her sons different surnames, Wang, Feng, and Shi, "in order to avoid the evil of conscription." It then recounts how a few generations later, the three Shi brothers Xian, Xue, and Cai moved to Ancestor House Mountain in what is now Shiyan village, cleared the forest, and started making paper. It then continues:

At that time, [the three brothers] established the name list for the coming twenty generations. For more than three hundred years their descendants have followed these rules, up to the present day. Our number has greatly increased to several tens of thousands, so that one can truly say we excel among the hundred surnames of the Chinese nation (*shi ke cheng Zhonghua minzu bai jiaxing nei zhi jingming ye*). Now we see that the list of twenty names laid down by our ancestors is gradually coming to an end. How could we, the mighty Shi family, strong in numbers and wealth, disgrace the sacred virtue of our ancestors by leaving the coming generations without terms for seniors and juniors? Therefore, many personages and the entire masses of our lineage, bearing in mind the benefits of the art of papermaking transmitted to us by our ancestors, with one heart and one mind discussed the matter of extending the name list, so that upwards, we can requite the teachings of the ancestors, and downwards, we can teach our descendants moral behavior and propagate the standards of loyalty, filial piety, propriety, and righteousness. At this time, when the country is peaceful and prosperous, in the second month of the *ren shen* year [1992], the masses unanimously began preparations and obtained full support of the authorities of the various villages. The broad masses vied with one another to donate funds of more than 10,000 *yuan* total. In the course of two years, we opened a quarry to cut the stone, dug out earth to lay a foundation, leveled the site, and carved the characters. Today the work is completed. All of us Shis shall obey, respect, and protect it.

> 1993, *gui you* year of the lunar calendar,
> on an auspicious date of the first month.

This is an awkward text, moving from stately classical Chinese— the "sacred virtue of our ancestors," "the standards of loyalty, filial piety, propriety, and righteousness"—to the bureaucratic jargon of state socialism—"full support of the various village authorities," "the broad masses," "the Chinese nation," and so on. Its main authors,

Shi Dingliang and Shi Dingsheng, said that the first section of the text was copied from an older stele in Mianzhupu. As I found out later, the Mianzhupu stele dates the arrival of the "first ancestor" to the Hongwu (1369–98) rather than to the Wanli (1572–1620) period and does not include a precise date for the move of the three Shi brothers to Jiajiang. Dingliang admitted that the true dates were not known but said that their amendments made the text more coherent and plausible. The values expressed in the text, he continued, were part of the "civilization of the Chinese nation" (Zhonghua minzu de wenming). Beifen, the distinction between junior and senior generations, was an old element in Chinese culture, conducive to social order and fully compatible with the political system of the PRC. If beifen had lost some of its force after 1949, it was because of poverty, not because the CCP opposed beifen. Now that living standards had improved, people no longer needed to worry about food and could afford to spend time on improving their spiritual life. Dingliang stressed that "not one word in the text was about politics" and that the Shis firmly opposed "feudal" customs such as physical punishment for infractions of lineage rules or government by lineage elders. The fact that they had gone through the proper channels to obtain support proved that they had no selfish political agenda. The Shis' action, moreover, was not directed against anybody. Neighboring kinship groups had shown great interest in their example, and Dingliang was optimistic that some of them would follow the Shis' lead.[5] Other people in the village (including a few non-Shis) were less defensive but agreed that this was about traditions and morality, not about politics, and that the Shis and their neighbors would benefit from a more united and civilized village.

What the Shis did was by no means unusual. Since the demise of the collectives, kinship groups all over rural China have rebuilt ancestral halls, written genealogies, and compiled generational name lists.[6] Such acts are often seen as attempts to fill a power vacuum that arose after decollectivization, but this does not appear to be the case in Shiyan. On the whole, the village was well governed, and the village committee firmly in command. There is no indication that the stele was in any way directed against the village leadership or aimed at establishing an alternative center of power. The village leaders remained in the background; after all, the building of the stele was an unofficial (*minjian*) action by a single kinship group, albeit one that accounted for 80 percent of the village population. As representatives of the party-state and the entire village, the party secretary and his deputies could not take the lead in the Shis' affairs; as private persons, they fully supported them.

This raises the question of what the Shis sought to achieve by erecting the stele. On one hand, the stele can be read as a charter, explicitly endorsed by all Shi households through their donation of funds and their consent to having their names inscribed in stone. It is part of a formal reconstitution of the Shis as a bounded group with a clearly named membership, some sort of internal structure, and a shared sense of origin and history. On the other hand, although the stele may have reinforced the Shis' sense of identity, it did not constitute them as a group capable of action. In contrast to some other kinship groups, the Shis did not give themselves a lineage council or any other institution that could have been used to lobby for influence or speak with a unified voice. All the Shis gave themselves was a name list that they hoped would unite their descendants for the next twenty generations—but whether this would happen was something they could neither influence nor know.

Although the authors of the stele text denied having a political agenda, it seems to me that they make a political argument, if only obliquely. The stele was built at a time when the Shis emerged from decades of poverty, caused by state policies that had turned them into peasants; a time also when they were reaping extraordinary benefits from the revival of papermaking. This is not directly discussed, but the inscription creates a strong link between agnatic identity and craft skill. The Shis, the stele text says, owe their present prosperity to the "sacred virtue of their ancestors"; to show their gratitude, they must transmit the "teachings of the ancestors" to future generations. These teachings include, first and foremost, the "standards of loyalty, filial piety, propriety, and righteousness," but they also include "the art of papermaking," which is the basis for their current wealth and strength. The text does not say that future generations have an obligation to become papermakers, but clearly papermaking is seen as part of an ancestral heritage that must be cherished and respected.[7] Clearly, the link between generational continuity and occupational identity remains strong in Shiyan. When asked why they make paper rather than look for factory jobs or migrate to the cities, people often refer to papermaking as an an-cestral tradition: "we Shis have always made paper" (*women Shi jiazu lilai caoguo zhi*) or "we have been making paper generation after gen-eration" (*women jiazu zuzu beibei caoguo zhi*). In fact, not all paper-makers in the village are Shis, nor do all Shis make paper, but for most Shis, kinship and occupational identity are linked—as they are for the neighboring Mas, Yangs, Zhangs, and Xiongs.

At the same time, the text creates a time frame for the Shis very different from the narrative of progress that dominates political dis-course in the PRC. In this dominant narrative, the Shis are doubly marked as backward: as craftspeople, they practice an outmoded form of production; as peasants, they need to catch up with more advanced urban people and industrial workers. The stele text does not openly confront this narrative, but it subverts it by placing the Shis into a much larger time frame. Taken together, the two name lists promise continuity over forty generations, or roughly a thou-sand years: from 1666, when the first name list was issued, to the twenty-sixth or twenty-seventh century, when the new list will ex-pire. Events in the text are dated by reign period ("the fifth year of Kangxi") or using the traditional sixty-year cycle ("*gui you* year"). The language is self-consciously archaic, using such terms as *sheji* (the altars of land and grain) instead of the modern *guojia* (nation). All this dissociates the Shis from a modern discourse of progress and development that denies their cultural or social distinctiveness and places them in the larger context of Chinese civilization, in which they can claim a meaningful place of their own.

Another theme in the stele text is the obligation of all Shis to maintain the "teachings of the ancestors" as a common heritage. Not long before the text was written, the Shis had collectively re-plenished and expanded a stock of common knowledge that had worn thin during the Maoist years. They had done so by freely shar-ing skills with their neighbors and relatives and by cooperating across household borders. In the 1990s, however, workshops that had been open to their neighbors began to erect figurative and physical walls. Early household division and the growing unwilling-ness to cooperate across generations were socially disruptive; the rise of the small husband-and-wife workshop threatened the long-term survival of the industry. The fact that many Shis had gone into pa-per trading meant that Shi papermakers now sold their paper to other Shis, a source of conflicts that had not existed in the past. One way to read the stele, then, is as an appeal to the Shis to ensure that the benefits from papermaking were shared equally and equitably. The emphasis is not so much on "this is *our* property" as on "this is our *common* property": the ancestral art of papermaking belongs to all descendants and must not be monopolized by the wealthy few.

Preaching *Beifen* to the Relatives

In 1998, I had the opportunity to listen to Shi Dingliang ex-plain these themes to his distant lineage cousins during a trip to Mianzhupu. Shi Dingliang, it will be recalled, was the former land re-form activist and commune party secretary who co-authored the stele inscription. While drafting the text, he repeatedly visited the ances-tral home of the Wangs, Fengs, and Shis in Mianzhupu, where the three surnames used to have their joint lineage hall. The hall itself had been destroyed during the Cultural Revolution, but Dingliang said that a stele that documented the early history of the three sur-names had survived. On our visit to Mianzhupu we found the stele, although since Dingliang's last visit it had been used to cover a cess-pool and its face had been partly obliterated by ammonium fumes. Af-ter copying the legible portions, we began to look for elderly villagers who could tell us more about naming practices in Mianzhupu. Since the variant of Sichuan dialect spoken in Mianzhupu was very differ-ent from the one used in Jiajiang, I left the talking to Shi Dingliang. During our walk through the village, Shi Dingliang introduced him-self as lineage cousin from Macun and me as a friend interested in the history of the Shis. Despite its proximity to the tourist hot spot of Leshan (home to a giant Buddha statue), Mianzhupu had seen few foreign visitors, and the people we talked to were evidently confused by a person who spoke some Chinese but looked foreign. When they asked Dingliang if I were also a Shi or a relative, he explained that I was a *waishengren*, a person from a different province, and thus avoided labeling me as a *waiguoren*, or foreigner.

As we were looking for the person who had provided Dingliang with information during his earlier visits, a small group gathered around and kept us company as we walked through the village. We soon found out that Mianzhupu was still inhabited by the Wangs, Fengs, and Shis, as well as other surname groups that had joined them later. The Fengs, with 200–300 households, were the most numerous, followed by about 100 Wangs. Only ten to fifteen Shi households were left. All three groups had used some sort of generational naming in the past, but their name lists (which were all different both from one another and from that of the Shiyan Shis) had fallen into disuse after 1949. Generational naming had survived longest among the Fengs, many of whom still remembered their "old twenty names." It was weakest among the Shis, who had either lost their common names or perhaps never had them. Some Shi families in Mianzhupu used lists of five generation names and started again from the top once the last name was reached.

It took us about an hour to establish these facts. By that time, my curiosity was satisfied, but Shi Dingliang was becoming increasingly exasperated with his lineage cousins' disregard for generational order. As a former commune party secretary, he knew how to capture the attention of his audience, and the crowd that had gathered around us grew as we went through the village. Both Dingliang and I were struck by the poverty of most of the homes we visited. Mianzhupu lies on the road from Chengdu to Leshan, in a fertile plain dotted with towns and village factories. However, the village had no industrial enterprises and depended entirely on farming. Mud-walled and straw-roofed houses, which by that time had disappeared from Shiyan, were still common, and some of the homes we saw were furnished with nothing but a bed and a few bamboo stools—in stark contrast to Shiyan, where most households had TV sets, tape recorders, and fully furnished bedrooms and living rooms. Dingliang linked the poverty of his cousins to their failure to practice *beifen*, and the greater prosperity of Shiyan to their strongly developed sense of agnatic solidarity. Although it seems just as likely that prosperity is the cause rather than the effect of lineage solidarity, the argument seemed to make sense to Dingliang's listeners.

Dingliang introduced himself to each new listener as "a Shi from Jiajiang," followed by the observation that "we Wangs, Fengs, and Shis are all brothers" and the self-depreciating statement that "we Shis are the youngest of the three families."[8] He then related how the Shis in Jiajiang were approaching the end of their name list and had recently decided on a new set of names, and recommended that the Wangs and Fengs do the same. After a first visit to an extremely poor Shi family, all other people we talked to were Wangs and Fengs, and Shi Dingliang pointed to the poverty and weakness of the remaining Shis to underline his message. In contrast to the Wangs and Fengs, many of whom still remember their former name lists, the Mianzhupu Shis lost their list long ago. By using different names in different families, they had lost their unity as an agnatic group. Even worse, by "recycling" the same names every five generations, they confused the generational hierarchy: "What good can come from a naming system that makes your children your ancestors?" "If you mess up your name list, there will be problems. I am not talking about poverty; with messed-up names, you may become like your Mianzhupu Shis. Just look at Shi Wanfu!" This referred to an exceptionally poor family whose adult son had lost a leg, apparently in an accident. The defect was neither congenital nor mental, but Dingliang went on to argue that there was a scientifically proven connection between generational order and the health of one's offspring: "If people of the same family intermarry, their children will be mentally deficient (*naoke hui chu maobing*). This isn't something I have invented; it is proven by scientists." One of the aims of extending the name list among the Shis, Dingliang said, was to prevent a situation in which agnates could no longer identify one another as belonging to the same descent group and would thus intermarry. The Chinese Marriage Law forbids marriage between collateral relatives by blood up to the third degree of relationship, but this, according to Dingliang, was only a minimum requirement. When he and his fellow Shis proscribed marriage between *all* agnatic relatives, they amended rather than contravened the Marriage Law.

In his remarks, Dingliang drew on the official discourse on eugenics and "population quality" (*renkou suzhi*). China is the only major state worldwide that is actively trying to improve the genetic makeup of its population—especially of rural people, who are accused of a lack of reproductive discipline, inbreeding, and inferior genetic quality.[9] Eugenics is an integral part of the one-child policy, which aims to ensure not only "fewer" and "later" but also "better" births. Rural people are generally familiar with the notion that marriage between close kin can lead to congenital diseases, but there is little understanding of Mendelian genetics and a general ignorance as to what kinds of relationships are likely to produce deformed births. This made it easy for Dingliang to appropriate the state's eugenic discourse and shift the emphasis from marriage between genetically close kin to marriage within the agnatic kinship group. In genetic terms, Dingliang's argument makes little sense: *beifen* is exclusively concerned with order between agnatic kin, who in any case do not intermarry. The former taboo against marriages between the Wangs, Fengs, and Shis is no longer strictly observed, although such marriages remain rare. However, Dingliang specifically referred to marriages between people of the same surname (*tongxing*), which quite simply never occur: Wangs in Mianzhupu do not marry Wangs, Fengs do not marry Fengs, Shis do not marry Shis. If there is any danger of congenital defects, it comes from the fact that men marry their cross-cousins (father's sister's or mother's brother's daughters) or matrilateral parallel cousins (mother's sister's daughters).[10] In all these cases, the cousins have different surnames and are therefore seen as belonging to different kinship groups. Such marriages were seen as unproblematic and even desirable; they were legal until 1980 and relatively common among the Jiajiang Shis.

Marriage proscriptions under Chinese law are ostensibly based on modern science and are similar to those of other nations; no concessions are made to traditional agnatic ideology. However, modern Chinese conceptions of race, eugenics, and sexual hygiene grew out of an earlier concern with the maintenance of a well-ordered agnatic body, and at least at the popular level, the two are easily confused. Generational order was traditionally associated with prosperity and health, and confusion in the generational ranks with decay and decline. The Qing code, not fully replaced until 1930, punished unions between persons of different generations, even if they were not related by blood (e.g., a son and his father's concubine). Popular sentiment against such unions ran "so high that even a teacher marrying his or her pupil, or a person marrying a friend's daughter or son, [was] condemned."[11] Similarly, adoption of agnates from the next-junior generation was approved and widely practiced; adoption of a relative belonging to any other generation could be punished with 60 strokes of the heavy bamboo. All these acts could be classified as *luanlun* (confusion of human relations), a term that today means incest but more generally referred to transgressions of generational boundaries. For late Qing intellectuals such as Yan Fu and Liang Qichao, the lineage (*zu*) served as a model for the nation or race (*minzu*). It was from this "concern within families over the continuation of the lineage that human sexuality and reproduction gradually emerged as a public domain linked to the strength of the country."[12] As a vast literature on reproductive health and "fetal education" shows, the production of healthy offspring was a central concern for Qing and Republican elites and has once again become so in the post-Mao period. A common theme in this literature is that order and harmony improve the "quality" of one's offspring, whereas social, moral, or aesthetic disorder (*luan*) produce defective births. The ease with which Dingliang switched between *beifen* discourse and the official rhetoric suggests that popular and official discourses on reproduction remain intimately related.

A recurrent theme in Dingliang's discourse was the link between generational order and prosperity. Rather than arguing a causal link, Dingliang let the facts speak for themselves: the Shiyan Shis arrived in Jiajiang as migrants from Mianzhupu, and despite the remoteness and poverty of the hills, they are now wealthier and more numerous than their cousins in the plains. The reason, Dingliang suggested, was that they had put their house in order by using proper generational names. The Mianzhupu Shis, by contrast, had messed up the agnatic body by using a naming system that "made their children their ancestors" and were now suffering the inevitable consequences. This link between generational order and prosperity is intuitively convincing: people know from experience that prosperity and generational order often went together. A skeptic, of course, might argue that prosperity was not the consequence of generational order but rather its precondition. Some people in the audience seemed to think that way: generational names, they said, were a good thing, but they were impracticable in such small and weak groups as theirs.

Kinship, Occupation, and Identity in Post-Reform China

At one point, when a large crowd of Wangs and Fengs surrounded him, Shi Dingliang recalled how the first ancestors of the Wangs, Fengs, and Shis had come to Mianzhupu from Xiaogan township in Hubei. He then paused for a moment, and in the ensuing silence, an old woman could be heard to sigh "Ah, Xiaogan!" Xiaogan is the distant home to which the three surnames (like so many other Sichuanese) trace their origin; the place-name means "sentiment of filial piety." Dingliang's vision of generational order as a precondition for the good life held strong appeal for the elderly, who complain that in post-reform China "old people are servants and daughters-in-law are bosses" (*dang laoren shi dang yongren, xifur dang laoban*).[13]

Although in no way oppositional, Dingliang's message can be seen as subcultural, as the expression of values that enjoy little legitimacy in the eyes of state officials and urban people.[14] Such values have to compete not only with a state-backed ideology of household individualism that promotes nuclear families as independent, market-oriented, entrepreneurial cells, but also increasingly with an ideology of individualism *tout court* that celebrates the self-interested, acquisitive, hardworking economic actor.[15] As I have shown in the preceding chapters, individualism has become the dominant ideology in reform and post-reform Jiajiang—so dominant, in fact, that it compels people to reject forms of economic cooperation that make much economic sense. Yet unmitigated individualism is unlikely to hold much long-term appeal for any but the youngest, strongest, most successful actors—and not even necessarily for those.

Official ideology offers little that rural people can identify with. Chinese nationalism is implicitly anti-rural in its emphasis on "catching up with and overtaking" the nations of the West and overcoming China's backward past.[16] Beyond nationalism, rural people can have recourse to a Maoist rhetoric that celebrates peasants as the backbone of the revolution, but only at the expense of deepening the gap between them and an urban population that sees in such nostalgia a further proof of peasant backwardness. There are few intermediate groups or institutions that invite positive identification from rural people: counties, districts, and provinces are mainly administrative units with little emotional content, and identification with strangers on the basis of shared lifestyles and consumer tastes is the preserve of wealthy and well-educated urban people. In such a situation, kinship and village membership remain the strongest bases on which rural people can construct a collective identity. There is no guarantee, of course, that such identities will last: the Mianzhupu Fengs and Wangs were impressed by Shi Dingliang's plea for agnatic solidarity and correct naming practices, but they are unlikely to establish a new list of generation names. Nor is it at all certain that the Shiyan Shis will continue to practice generational naming for the next two or three, let alone the next twenty, generations. Yet for the time being, group formation on the basis of shared descent continues to appeal to rural people, as the widespread construction of lineage halls and the compilation of genealogies shows.[17] Such communities are not necessarily "for" anything in particular; they are realms of increased emotional density in which a large variety of transactions are conducted more easily than "in society" (*zai shehui*). Whether reviving *beifen* naming practices can check the tendency toward what Yan Yunxiang has called "uncivil individualism" remains to be seen.[18]

Conclusion

THE SMALL-HOLDING PEASANTS form a vast mass, the members of which live in similar conditions but without entering into manifold relations with one another. Their mode of production isolates them from one another instead of bringing them into mutual intercourse. . . . A small holding, a peasant and his family, alongside them another small holding, another peasant and another family. A few score of them make up a village; a few villages make up a Department. In this way, the great mass of the French nation is formed by single addition of homologous magnitudes, much as potatoes in a sack form a sack of potatoes.

This quotation from Marx's *Eighteenth Brumaire* sums up the nineteenth-century European view of peasants as living outside the division of social labor and therefore outside civil society. Marx's views on "the idiocy of rural life," "the fixity and sameness of conditions" that make the peasant "the barbarian in the midst of civilisation," were not unusual, nor were Marx and Engels particularly harsh in their judgment.[1] Until the end of the nineteenth century, educated Frenchmen described peasants as savages "full of hatred and suspicion," "hardly touched by civilisation," belonging to "another race"—despite the fact that peasants in post-revolutionary France were better off economically and had more rights than peasants in most other parts of Europe.[2] The transformation of peasants into citizens of nation-states is largely a modern phenomenon, driven by the dual logic of capitalist development (with its need for formally free labor) and state-making (with citizen rights a trade-off for increased taxation and conscription).

Most theories of citizenship describe a process of gradual expansion, in which individual liberties (foremost the individual right to property) were first won by the urban bourgeoisie in the revolutions of the eighteenth century and were then step by step extended to the less privileged classes. As access to citizenship widened, citizenship rights deepened to include not only civil and political but also social rights.[3] Recently, Margaret Somers has proposed a less linear and teleological reading of the formation of citizenship rights in the archetypal English case. She contrasts a Lockean tradition that sees all liberty as rooted in rights to property produced through autonomous labor with an older, partly submerged tradition of "relational" rights, contingent on membership in self-regulating and self-perpetuating communities, groups, and networks. In medieval guilds, "free" (nonmanorial) villages, and pastoral or proto-industrial communities, individuals were not primarily bearers of exclusive individual rights but *members* who could make claims on other members.[4] Membership in such communities gave access to positive rights and redistributive justice, in the form of regulated employment, guaranteed livelihood, and "mutual aid, religious life, social organizations—indeed an entire cradle-to-grave culture."[5] Membership was also the primary means through which people enjoyed legal protection and political representation. In guilds and proto-industrial villages (which often developed guild-like structures), membership was acquired through apprenticeship—a process that served to induce novices into the "mistery" of the craft, in the dual sense of technical skill (as in "the mistery of weaving") and as the medieval term for a guild or guild-like body.[6] The "property of skill," in Somers' formulation, was primarily a social membership, a precondition for inclusion in a particular fellowship or body and, through this body, in the wider polity. Although Somers does not elaborate on this, it is clear from her account that such communities were essentially communities of practice, that is, communities united by the pursuit of a common livelihood and the shared use of material and immaterial resources.

China, of course, has no native language of citizenship rights, but it does have a long tradition of localized guilds, village councils, religious associations, and other groups that regulate themselves and represent their members vis-à-vis state agents or other groups. Timothy Brook labeled this type of group formation "auto-organization," a process that "has gone on within Chinese society since the emergence of the state, often out of its sight, and sometimes in tandem with . . . state interventions at the local level."[7] Although Chinese states did not grant anything like formal citizen rights, membership in guilds, villages, and kinship groups did convey substantial rights, both in terms of access to communal resources and in terms of a rightful place within the wider community and culture. In contrast to Europe, where for a long time the word "citizen" meant "simply inhabitants of a town, in sharp contrast to those who were subjects under feudal princes and kings in the countryside," China knew few legal and administrative barriers between town and country.[8] Although there is plenty of evidence that city dwellers thought of rural people as ignorant yokels, David Faure and Tao Tao Liu are probably right in arguing that "in the Ming and Qing dynasties, rural-urban distinctions were not a significant part of an individual's identity. Only in the early 1900s, when political reforms separated cities and towns as agents of social change, did an ideology emerge that looked upon villages as sources of backwardness."[9] Rural people were integrated into the polity (through households, numerical systems of registration, and the imperial bureaucracy), the society (through kinship, neighborhoods, guilds, and religious associations), and the economy (through market networks that spanned cities and villages) in much the same way as urbanites. Similar institutions in town and country facilitated economic interpenetration. Cities and villages formed part of a patchwork of functionally differentiated, interdependent places, with their own local culture, their local products, and their economic specializations.

The nexus between local origin and economic specialization remained important in the early stages of Chinese industrialization. Chinese factories were staffed not by generic proletarians but by occupationally specialized migrants—shipwrights from Guangdong, silk weavers from Hangzhou, factory hands from Subei—who brought their work experiences and workplace cultures to the cities where they sojourned.[10] These cultures and experiences informed workers' views of what it meant to be a worker, what constituted "fair pay for fair work," what their rights and obligations were vis-à-vis other workers and employers. As Elizabeth Perry has shown, the politics of

Chinese labor were to a substantial degree "politics of place"—place in the dual sense of local origin and of social status, membership, and belonging. Place in the sense of local origin was at the heart of workers' identities, since workers defined themselves in terms of geographical origin rather than class. Place in the sense of membership and belonging shaped working-class politics, since workers came from locally specific "cultures of attachment" in which each individual was tied into a complex matrix of rights and obligations and struggled to create similar forms of reciprocity in the new workplace.[11] Perry convincingly argues that one of the central institutions of socialist China, the urban work unit (*danwei*) with its permanent membership and cradle-to-grave welfare provisions, grew out of an artisanal culture characterized by guild regulation, substantive rights to welfare, livelihood, and employment, and an emphasis on reciprocity and mutual obligation.[12]

Although "place" is a useful shorthand term, it is not origin or residence as such that underlies such cultures of attachment. Groups like the Jiajiang papermakers are defined by multiple, loosely overlapping categories: residence, occupation, skill, descent, and membership in formally constituted guilds or religious associations. All these are important, but I believe such groups are best understood as communities of practice, united by the pursuit of a common enterprise (often sustained over generations) and the joint use and management of communal resources, including a common stock of skills, experience, and knowledge. It was on this basis—as an internally structured group of co-practitioners, linked to one another and differentiated from other groups by the pursuit of a common trade— that Jiajiang papermakers were integrated into the economy, polity, society, and culture of late imperial China. As in the guilds and proto-industrial communities discussed by Somers, particular local memberships and attachments were not antithetical to inclusion in a wider universe; to the contrary, it was the only meaningful way for local people to participate in a larger whole. Particularism became a problem only when modernizing governments insisted on direct, unmediated access to the population. As in post-revolutionary France, where "diversity became imperfection, injustice, failure, something to be noted and to be remedied," particular identities and attachments in post-revolutionary China were suppressed in the interest of the greater unity of the nation and the people.[13] At the same time, specific differences that had not been marked in the past were now perceived as natural and necessary, and were consequently enlarged and uniformly imposed. People became rural or urban, *nongmin* or *gongren*, closely tied to one of these mutually exclusive spheres through class status, household registration, and affiliation with a rural collective or urban work unit. Some of these categories were innovations and did not reflect any pre-existent reality; in other cases, previously fluid distinctions hardened into boundaries. The outcome was, ironically, that papermakers now experienced the "fixity and sameness of conditions" that Marx had ascribed to peasants but that had been quite absent from their previous lives.

The issue here is not whether CCP policies were ideologically "anti-peasant." Marx's and Lenin's contempt for the "backward" peasantry was never shared to the same degree by the Chinese Communists, who glorified peasants as morally superior while denying that they could be independent, self-conscious historical agents. As James Scott has shown, modernist governments of very different ideological persuasions were united in their distrust of historically grown, complex, and therefore opaque structures, which they aimed to replace by a neat and transparent order. In the view of Chinese state planners, modernity lay in the rational division of the economy into discrete, hierarchically ordered jurisdictions, arranged and interlinked in such a way that a person at the top could trace and direct the flow of resources through the system. "Auto-organization," whether economic or social, was suspect—especially if it worked—because the success of grown structures threatened the perceived higher rationality of the plan.

Seeing Like a State, Smelling Like a Market

In an astute and entertaining review of Scott's *Seeing Like a State*, Fernando Coronil asks us to consider not only "how states see" but also "how markets smell"—smell in the dual sense of emitting stenches or fragrances and of "smelling out" successful products, opportunities, or consumer needs.[14] By focusing on the authoritarian state, Coronil asserts, Scott overlooks the role of capitalist markets as agents of high modernist designs and the historical and ongoing collusion between capitalist markets and modernizing states. Scott himself avers that "large-scale capitalism is just as much an agency of homogenization, uniformity, grids, and heroic simplification as the state is, with the difference being that, for capitalists, simplification must pay." Coronil wholeheartedly agrees but takes Scott to task for a critique of the "really existing market" that is so thin and abstract that he ends up lending support to dominant neoliberal models. Not only does Scott locate homogenizing and objectifying tendencies solely in the state, but he also contrasts the state with an improbable world of small property-owning artisans and peasant farmers unhampered by institutional constraints. In doing so, he underwrites a dualistic view in which society, and by extension the market, "is the locus of individuality and common sense in opposition to the state as the domain of authoritarian practices and impractical designs."[15]

My focus, like Scott's, is on state visions and their consequences. Like the modernizing states in Scott's narrative, Maoist China preferred "bulldozed sites" (or, to use Mao's formulation, "blank sheets of paper") to historically grown structures. Rural craft industries of the type found in Jiajiang were seen as a form of industrial wild growth that needed to be weeded out and replaced by economic forms more closely corresponding to the imagined rational economic order. It is striking how much PRC policies in Jiajiang were guided by a desire for symmetry and order, for neat divisions between sectors, for visibility and control, rather than by considerations of productivity, profitability, and efficiency. This desire for pure forms and the concomitant fear of messiness led to deindustrialization and famine in the Jiajiang hills, the loss of untold lives, a precipitous drop in living standards, and the destruction of the social and cultural tissue that had linked papermakers to the state and society at large. Markets, by contrast, were not a threat to Jiajiang papermakers. Paper production flourished whenever producers had access to grain and paper markets and suffered when their access was interrupted by grain shortages, war, or state regulations. Yet my argument here is not that market forces are always benign. The experience of other groups of specialized rural producers was the exact reverse of what I found in Jiajiang: Eric Mueggler, for example, describes how a community of (non-Han Chinese) Yi people in a remote part of Yunnan retained a way of life based on hemp cultivation and handloom weaving throughout the collective period. In the 1980s, state trading organs were dismantled or withdrew from the market, and the hemp gunnysacks that had been the mainstay of the local economy were replaced by nylon bags. In this case, people experienced the post-Mao era as an "age of wild ghosts," characterized by the loss of their accustomed way of life, a drop in living standards, and a feeling of anomie and isolation—in other words, much in the same way as Jiajiang papermakers experienced life under the collectives.[16] Such cases must have been common in the reform years, as plastic, nylon, vinyl, concrete, and mass-produced textiles superseded handmade bamboo objects, straw mats, hemp ropes, wooden furniture, hand-formed bricks, ceramics, and handloom textiles, and rendered obsolete the ways of work and life that had evolved around them.

How are we to compare the homogenization imposed in top-down fashion by a powerful state committed to a reductive vision of rationality and modernity with the homogenization resulting from impersonal and anonymous market forces? How are we to conceptualize deskilling, deindustrialization, and other forms of "knowledge expropriation" under socialism and under capitalism? One obvious difference is that, in Scott's words, "for capitalists, simplification must pay." Despite the frequent attribution of a will to power to individual capitalists and to (global) capitalism as a system, what counts for capitalists is profit, not power. Historically, capitalists have sought to break workers' hold over the production process whenever this was technologically feasible and, above all, whenever it was profitable; when it was not (and often it was not—controlling workers can be very expensive, and a self-motivated and self-disciplined workforce may be cheaper and more efficient than a deskilled one), they have refrained from doing so.[17] Deskilling under capitalism follows two different trajectories. First, controls are imposed over the labor process through the separation of management and design from mere execution and the fragmentation of complex, skilled operations into the simple, unskilled ones. Second, skills are lost because entire industries are superseded when they become technologically obsolete—a fate that Mueggler's hemp producers share with millions of other small-scale handicraft producers in China and elsewhere in the world. Blanket deskilling of entire populations in the name of a higher rationality makes no sense as a capitalist strategy. In fact, although there is clear evidence for ongoing deskilling under capitalism, there is little evidence that total skill levels are declining.[18] In formal terms, skill levels are increasing in most parts of the world, as more and more jobs require literacy, numeracy, and advanced technical training. Whether this actually amounts to skilling hinges on the question whether formal qualifications translate into increased autonomy and control over work. It is easy to imagine highly qualified work (for example, in large banks or software companies) that takes place under such stringent managerial control that it is actually experienced as deskilled.

What deskilling under capitalism lacks in breadth, it makes up for in depth. Paradoxically, unplanned, uncoordinated market forces penetrate more deeply into people's working life than even the most ambitious and authoritarian state. Socialist states and other aggressive modernizers—"latecomer states" of liberal, militarist, or fascist persuasions; colonial administrations; post-colonial regimes—have sought to bring their national economies up to speed by banning or forcibly transforming the "backward" practices of farmers and small artisanal producers.[19] They have, as Scott amply demonstrates, wreaked havoc with existing structures, but they have rarely succeeded in achieving a lasting transformation. This is in part because even the most aggressively transformative states find it difficult to change social and technological relations at the point of production. This is true even for socialist states, which are ideologically committed to the view that social relations are ultimately determined by how value is produced and extracted in the workplace. Since they are not directly concerned with profit, states have no reason to do what industrial capitalists routinely and necessarily do: intervene directly in the concrete details of the production process. To use a distinction suggested by Michael Burawoy, states may be concerned with relations *of* production (the relations through which goods and services are appropriated and distributed), but they are seldom concerned to any significant extent with relations *in* production (the relations that describe the production of these goods and services).[20]

States in Sichuan became involved with relations *in* production twice, during the Sino-Japanese war and in the 1950s. In both cases, a dire need for paper and the impossibility of developing mechanized papermaking in the time available induced the state to turn to manual papermaking and to try to understand production technology, as a necessary precondition for controlling the production process. The Republican state never came close to succeeding, because people in the Jiajiang hills resented and resisted state attempts to wrest technological control from their hands. By contrast, the socialist state succeeded to an amazing degree in mapping and describing the production process. But when it came to putting this knowledge to practical use, PRC cadres failed almost as dismally as their Republican predecessors. Light Industry Bureau technicians and administrators in Jiajiang—on the whole well-intentioned and capable—developed simple, useful technologies that were well adapted to the local situation and were, in fact, enthusiastically endorsed by papermakers—but only after a delay of almost thirty years and the demise of the collective system.

The failure of the collectives to transform production technology is overdetermined. Several factors—the state's obsession (which turned into a real, pressing need after the Great Leap Forward) with maximizing grain production, the ideological strictures against a form of industrialization that did not directly "serve agriculture," the absence of the profit motive, the need of team leaders to create jobs even if this lowered productivity—are each of them sufficient to explain why paper production remained technologically stagnant under the collectives. Another and perhaps less obvious factor deserves fuller consideration: the inability of state agents (as institutional actors, though not necessarily as individuals) to come to terms with the nature of physically embodied, socially embedded skills. At all levels of the hierarchy, the reigning model was that of the factory or workshop as a mechanism composed of inanimate parts, governed in top-down fashion by a single intelligence. Although the Light Industry Bureau occasionally called for higher pay for skilled workers or for the preservation of rare skills (often when it was too late—

there are repeated references to embittered old masters taking their secrets to the grave), there is no real recognition of skill as a crucial resource that can and should be nurtured. Even less is there a recognition of a historically grown production structure as a form of distributed intelligence, a reservoir of useful knowledge, attitudes, and dispositions. Most paradoxically, given Mao's emphasis on the wisdom and creativity of the working masses and his insistence that all correct understanding comes from concrete practice (the ideal type of which is the transformation of matter through productive work), there was absolutely no recognition that groups of practitioners might be capable not only of reproducing old knowledge and skills but also of generating new knowledge. Historians of technology in Europe and Japan have linked technological creativity to institutional clusters that involve small family firms, intermediate organizations (guilds, local councils, kinship networks), and local traditions of informal cooperation, mutual aid, and knowledge sharing, often located in areas with strong artisanal traditions.[21] None of this was apparent to state planners, who tended to see papermakers simply as labor power, and who seemed incapable of seeing that local stocks of knowledge could be as much a locational advantage as soft water or abundant fiber supplies.

The inattention of state officials to the details of skilled work meant that the actual labor process remained untransformed. After the attempt in the 1950s to extract useful knowledge from the papermakers and to transcribe it and put it in circulation, state agencies withdrew almost totally from the papermakers' workplace. State policies toward the paper industry were determined by a tug-of-war between those parts of the administration that acted as patrons to the paper industry (above all the Industry Bureau/Handicraft Bureau/Light Industry administration) and those concerned about "illegitimate" grain consumption, a tug-of-war that resulted in the loosening or tightening of grain supplies to papermakers and in fluctuations in the price of paper. Apart from manipulating inputs and outputs (which had drastic consequences for the life and work experience of papermakers), state cadres rarely intervened in the paper industry. The same seems to be true for much of the Chinese countryside, where industries not phased out in the 1950s and 1960s often survived more or less unchanged into the reform period. Depending on luck, administrative patronage, and availability of grain, similar industries underwent dissimilar fates, leading to the creation of a patchwork of uneven technologies. Technological unevenness may have empowered workers in the surviving rural industries, who knew far more about how to produce their goods than their administrative principals.

Rural deskilling, then, remained incomplete, as pockets of skilled production survived in the countryside and new skills continued to emerge, sometimes thanks to, but more often despite, state policies. The inhabitants of Bishan village near Shiyan made paper until most teams were declared to be (and thus forced to become) "grain self-sufficient." At some point in the 1970s, some teams started to produce bricks and roof tiles from local clay. In other teams, women began raising silkworms. In Jinhua, another former papermaking village, tea cultivation and processing made up for the lost income from the paper industry. Elsewhere in the county, people started to make paper money for religious ceremonies, to produce contraband cigarettes, to process traditional Chinese medicine, and to quarry sandstone. Some of these sidelines were encouraged by the state, but most developed behind the back of the authorities. All of them require considerable skill.

Although skill levels remain high in many parts of rural China, this fact finds no social recognition. Rural people themselves are likely to react to the suggestion that they have skills by saying, "Oh no, we're just peasants."[22] State media and public discourse in the PRC characterize rural people as deficient in *suzhi*, or "quality," a term defined in a curiously circular fashion as the positive cultural achievements that China's rural people lack. In an article on *suzhi* and the representation of Chinese migrant workers, Yan Hairong has shown how accumulation in the post-reform period relies on the erasure of migrants' contributions. The people who build, enrich, and sustain China's cities and whose labor underpins the lifestyles of the urban middle classes are constructed as deficient: members of an undifferentiated peasantry who through their sheer numbers hold China back. Such stereotyping allows those who benefit from the exploitation of rural labor to construe exploitation as a "gift of development" in which the migrant worker is a recipient rather than a producer of net worth.[23] Managers and employers treat rural workers as peasants with "rough hands and rough feet" who handle precision tools as if they were about to plow a furrow.[24] In reality, of course, these are often dexterous and dedicated workers, coming from a rural culture in which skill-intensive production is a normal part of life. Rural workers are thus simultaneously *skillful* (the famous "nimble fingers"—evoked by the same managers that disparage their workers as "red and lazy") and *unskilled*, unable to translate actual technical skills into social status and material rewards. It is this disjunction that renders rural Chinese labor both productive and cheap and has buttressed much of China's recent economic growth.

Enclosure of the Commons

The way of life described in these pages is built on a combination of intense openness to markets and the existence of what is perhaps best described as a "commons"—a common stock of knowledge on which members can draw. This combination cuts across Western concepts that associate reciprocity and communal use of resources with simple, pre-capitalist societies, not with involvement in complex and efficient markets. There is, however, no inherent contradiction between the two, and the combination served Jiajiang papermakers well. There is no need to idealize communal bonds in Jiajiang. The fact that most of them were based on agnatic kinship meant that women had only the most limited membership rights. Women were consciously excluded, not from the knowledge of certain production processes (this was not technically feasible) but from regarding their knowledge as personal property that they could transmit at will. Skill reproduction was backed by gender and generational hierarchies that gave senior men the power to discipline and punish juniors. At the same time, *beifen* and the forms of reciprocity and mutual obligation that went with it may have slowed class differentiation and mitigated its effects. In Shiyan and other Hedong villages (though not always in the harsher environment of Hexi), even poor people could make claims on relatives and neighbors to acquire papermaking skills, to find work in paper workshops, to use temporarily unused equipment, or to find a sponsor who would sell their paper in his name. Communal investment in a set of technologies, ritually expressed as a shared obligation to the ancestors and reinforced on a daily basis by routine transactions, made sure that most people in the paper districts lived relatively secure lives.

This system of mutual obligations was remarkably resilient. It survived in a permutated form under the collectives and underpinned the rapid recovery of the paper industry in the 1980s. For most of the twentieth century, it remained outside the purview of the state. The "feudal" authority of lineage councils and generational elders was broken in the early 1950s, but this resulted only in a strengthening of the horizontal aspects of the kinship system. In Shiyan, where there is no history of lineage conflicts, where village leadership was stable throughout the collective years, and where political campaigns left fewer scars than in many other parts of China, reciprocal bonds remained relatively intact until the 1990s, as witnessed by the effort to enshrine this collective ethos in the Jiadangqiao stele. Yet the Jiadangqiao stele can also be seen as a sign that old certainties were wearing thin. At the time when the stele was built, technological change and a state-backed discourse of small-family entrepreneurialism had led to the emergence of small, barely sustainable workshops, increasingly dependent not on mutual help but on specialized service providers. Soon thereafter, the first mechanized paper workshop appeared in the village, a technological change that prompted the building of walls around household workshops, first only around mechanized workshops, then also around other large workshops. For the first time in its history, it appears possible that the technological commons that sustained the paper industry will be enclosed.

Appendixes

A   Character List for Selected Chinese Names and Terms

Entries are ordered alphabetically letter by letter, ignoring word and syllable breaks, with the exception of personal names, which are ordered first by surname and then by given name. For the names of varieties of paper and terms relating to papermaking technology, see Appendix B.

an 庵

ba da gongkou 八大公口
baijiazi 敗家子
baijiu 白酒
bai shi 拜師
Baizhibang 白紙幫
banche 板車
bang 幫
bangke, bang lao'er 棒客, 棒老二
bannong banzhi 半農半紙
baochan daohu 包產到戶
baochan daozu 包產到組
baojia 保甲
baoshou 保守
bao wo fuyuan 保我富源
baozhang 保長
Baxian county 巴縣

beifen 輩分
Beilin bowuguan 碑林博物館
beixu 碑叙
bendi fanyunshang 本地販運商
ben mo 本末
benren jieji chengfen 本人階級成份
biao 裱
Bishan village 碧山村
buqiu gailiang 不求改良
bu xiaoshun 不孝順
bu yao you zibeigan 不要有自卑感
Buzheng shi si 布政使司
bu zhichi, bu fandui 不支持, 不反對

235

caigouyuan 採購員
Cai Lun hui 蔡倫會
Cai Lun xian shi 蔡倫先師
Cao Shixing 曹世興
caohu 槽戶
Caohu daibiaohui 槽戶
  代表會
cha? cha ge pi! Dangshi cha bu
  chulai 查? 查個屁! 當時
  查不出來!
changgong 長工
chanpinshui 產品稅
chengbao 承包
Chengdu 成都
chi daguofan 喫大鍋飯
chi kuixinliang 喫虧心糧
chi Qingminghui 喫清明會
Chongqing 重慶
chuantong 傳統
Chuanxi xingzheng gongshu
  gongyeting 川西行政
  公署工業廳
Chuanzhu 川主
chuji shengchan hezuoshi
  初級生產合作社
chu'na 出納
chun zaozhihu 純造紙戶
citang 祠堂
cun 村
cuowu 錯誤

daba 大爸
dadan de gan 大膽地幹
dadui 大隊
da gong 打工
da houfang 大後方
dahu 大戶

dang laoren shi dang yongren,
  xifur dang laoban
  當老人是當傭人, 媳婦兒
  當老板
dangtian wang 當天王
daniang 大娘
Danling county 丹陵縣
danwei 單位
danzi dade 膽子大的
danzi xiaode 膽子小的
daobian zhengce 刀鞭
  政策
daocha men 倒插门
dapo cixiang louxi 打破
  此項陋習
da qing, xiao qing 大慶,
  小慶
daye 大爺
Dazhu county 大竹縣
dazibao 大字報
Deyang county 德陽
diaocha baogao 調查報告
dingdian xiangmu 定點
  項目
ding'eshui 定額稅
dixiong 弟兄
Dongyue 東岳
doufu 豆腐
duangong 短工
duilian 對聯
duizhan 堆棧
duobazi 舵把子
Duyun county 都勻縣

Emei county 峨眉縣
erba 二爸
erniang 二娘

Erqing gongye ju　二輕工
　業局

Fanggou village　坊沟村
fanggu　仿古
fangong jiuguo jun　反共
　救國軍
fanxiao　返銷
feinongye renkou　非農業人口
feishui buyao wailiu　肥水不要
　外流
fen jia　分家
fen jia bu qing　分家不清
fubaozhang　副保長
fu er yao jin　富而要進
Fu Lu Cai shen　福禄財神
Fuyang county　富陽縣
fuye zaozhi　副業造紙

gailiang　改良
gan huo　幹活
gan qin　乾親
gaoji shengchan hezuoshi
　高級生產合作社
gelaohui　哥老會
getihu　個體户
Gezhi huibian　格致彙編
gongbi　工筆
gongnong fen le jia　工農
　分了家
gongnong hunheshe　工農
　混合社
gongren　工人
gongren jieji　工人階級
gongshangyezhe jian dizhu
　工商業者兼地主
gongshi　工食

gongyoubang　工友幫
gongzuo　工作
Guang'an county　廣安縣
guangrong　光榮
guanli　管理
Guanyin pusa　觀音菩薩
Gufosi　古佛寺
gui'erzi meipo　龜兒子媒婆
gui you　癸酉
gungun li　滾滾利
gunong　雇農
guoji　過繼
guyong　雇佣

haokan zai lianpan shang,
　haoshua zai banshanyao
　好看在臉盤上, 好耍在
　半山腰
Hechuan county　合川縣
Hedong　河東
hehuo　合夥
hejia nao　合家闹
hengbian　橫區
Hexi　河西
Hongchuan　洪川
Hongya county　洪雅縣
hongzhibang　紅紙帮
hongzhi zuofang　紅紙作坊
Hua Younian　華有年
Huang Yonghai　黄永海
huan wo he shan　還我河山
Huatou township　華頭鄉
Huayang county　華陽縣
hui　會
huishou　會首
hukou　户口
hukoubu　户口簿

hunshui　渾水
huomi　火米
huoshitang　伙食堂
huoshitang xiafang　伙食堂
　下放
huzhuzu　互助組

Jiadangqiao　加檔橋
jiafatai　家法台
jiage daogua　價格倒挂
Jiajiang county　夾江縣
jiamiao　家廟
jiang beifen　講輩分
Jiangpai　江派
jiangshouliang　獎售糧
jianlou　簡陋
Jiansheting　建設廳
Jianshe weiyuanhui　建設
　委員會
jiapu　家譜
jiashen　家神
jiating chushen　家庭出身
jiating jieji chengfen　家庭
　階級成份
jiaxun　家訓
jiazhang　甲長
jiedao　街道
jingjian chengzhen renkou
　精简城鎮人口
Jingxian county　經縣
Jinhua village　金華村
jinshanhui　禁山會
Jinsha River　金沙江
jishu　技術
jishu gongren　技術工人
juezhong　絕種

kao benben de jiu xiguan
　靠本本的舊習慣
ketou　磕头
ke xue kai hui liang deng
　chao cheng an xiang bang
　qian si xian yi ze ren
　shang yuan chang　科學
　開惠良登超呈安祥榜前
　思賢宜澤仁尚遠昌
Kong Guan ersheng　孔關
　二聖
kongke cun　空殼村
kouliangdi　口粮地
kuaikuai　塊塊
kufang　庫房
Kunming　昆明

Lanzhou　蘭州
laobanbang　老闆幫
laodong renmin　勞動人民
laogong　老公
laoye　老爺
Leshan (Jiading)　樂山
　(嘉定)
Li Jingquan　李景泉
Li Xiannian　李先念
Liang Qichao　梁啓超
liangben　糧本
Liangshan (Liangping)
　county　梁山 (梁平) 縣
Liangshan Mountains　涼山
lifu　力伕
lingfang　弱房
lingxing buneng hezuo　零星
　不能合作
Liu Min　劉敏

Liu Shaoqi　劉少奇
Liu Wenhui　劉文輝
Liu Xiang　劉湘
Longtuo township　龍沱鄉
luan kaiqin　亂開親
luanlun　亂倫
Luo Guojun　羅國鈞
Luzhou county　瀘州

Ma Weishan　馬爲善
Macheng county　麻城縣
Macun township　馬村鄉
mai gong de　賣工的
mai qingmiao　買青苗
Maliu township　麻柳鄉
Meishan county　眉山縣
menglong　朦朧
menshen　門神
menshi　門市
Mianzhupu　綿竹鋪
miao　廟
miaohui　廟會
mi fanzi　米販子
minjian　民間
Min River　岷江
minzhuhua, helihua, qiyehua
　民主化, 合理化, 企業化
Mucheng township　木城鄉

Nan'an township　南安鄉
naoke hui chu maobing　腦殼
　會出毛病
Neijiang　內江
nianhua　年畫
Nonghui　農會
nongmin　農民

paihang zi　排行字
paoge　袍哥
pao shichang　跑市場
pao waixian　跑外綫
paoxuan　炮選
Peng Shaonong　彭少農
pengmin　棚民
pianpi　偏僻
piaoxuan　票選
pigu zuo wai　屁股坐歪
pinxiazhongnong　貧下中農
Puge county　普格縣
Pujiang county　浦江縣

qi　氣
Qi Baishi　齊白石
Qian'an county　千安縣
qiaoqiao li　敲敲利
Qiliping village　七里平
qingfei fanba　清匪反霸
Qing gongye ju　輕工業局
Qingminghui　清明會
qing ren　請人
qingshui　清水
Qingyi River　青衣江
Qiongzhou county　邛州縣
Quanguo jingji weiyuanhui
　全國經濟委員會
quanxiang　全鄉

ranzhifang　染紙坊
Ren Zhijun　任治鈞
renao　熱鬧
renkou suzhi　人口素質
ren yi li zhi xin　仁義禮智信
Rong county　榮縣

ruzhui　入贅

sanci daomei　三次倒霉
sanqiong sanfu bu dao lao
　三窮三富不到老
sanqiong sanfu huo dao lao
　三窮三富活到老
sanshi nian hedong, sanshi
　nian hexi　三十年河东,
　三十年河西
san-sishi nian de laopo　三四
　十年的老婆
shang ban　上班
shan'ge　山歌
Shanghui　商會
sheji　社稷
shenfubang　神袱幫
shengchan dadui　生產大隊
shengchan dui　生產隊
shenshang yao ganjing　身上
　要乾淨
shenti suzhi, sixiang daode suzhi
　身体素質,思想道德素質
Shi Guoliang　石国梁
Shi Longting　石龍庭
Shi Ziqing　石子青
shidian　試點
Shijiao village　石窖村
shi ke cheng Zhonghua minzu
　bai jiaxing nei zhi jingming
　ye　實可稱中華民族百
　家姓內之精明也
shilang　侍郎
Shinian village　石埝村
Shiyan village　石堰村
Shougongye guanliju　手工業
　管理局

Shougongye hezuoshe　手工
　業合作社
shougongye sanshiwu tiao
　手工業三十五條
Shuangfu township　雙富鄉
shuashua zhi　耍耍紙
shuitu　水土
si　寺
Sichuan haozi　四川耗子
Sichuan jun　四川軍
sida faming　四大發明
Su Hanxiang　蘇漢湘
suzhi　素質

Taiyuan　太原
Tangbian village　塘邊村
teda dizhu　特大地主
tian di　天地
tiaotiao　條條
tiezadian　貼扎店
tongbeizi　同輩子
tonggouhu　統購戶
tonggou tongxiao　統購
　統銷
Tongliang county　銅梁縣
tongxiaohu　統銷戶
tongxing　同姓
tongyangxi　童養媳
Tongye gonghui　同業公會
tougong jianliao　偷工減料
tu　土
tubu　土布
tudi shi jitide, zhulin shi jitide,
　shenghuo ge guan gede
　土地是集体的, 竹林是
　集体的, 生活各管各的
tufei wozi　土匪窩子

waiguoren　外國人
wailai caigoushang　外來
　　採購商
waishengren　外省人
waixing　外姓
Wang Feng Shi shi lidai zuxian
　　汪冯石氏歷代祖先
Wang Yunming　王云明
Wangpai　王牌
wanhui liquan　挽回利權
Wanxian　萬縣
weibei paihang　違背排行
Weiyuan county　威遠縣
wenfang sibao　文房四寶
wenren　文人
wenwei juanzhi　文圍卷紙
wen wu fuzi　文武夫子
women jiazu zuzu beibei caoguo
　　zhi　我们家族祖祖辈辈
　　操過紙
women Shi jiazu lilai caoguo zhi
　　我們石家族歷來操過紙
women zhefang de ren chi le
　　huikou　我們這方的人
　　吃了回扣
Wu Zuozhang　吳作章
wuchi zhi tu　無恥之徒
wuren shouhuotai　無人售貨台
Wuziju　物资局

Xi'an　西安
xiangdenghui　香燈會
xiangzhen qiye　鄉鎮企業
xianhuo　現貨
Xian, Xue, and Cai　賢, 學, 才
xiaobeizi　小輩子
xiao er quan　小而全

Xiaogan township　孝感鄉
xiaozu　小組
xibu cidu　西部瓷都
Xie Rongchang　謝榮昌
Xi'nanqu gongyebu　西南區
　　工業部
Xinghuo factory　星火廠
xinyong　信用
xiucai　秀才
X shi tangshang lidai xianzu
　　kaobi zhi shenwei
　　X氏堂上歷代先祖考妣
　　之神位
Xu Beihong　徐悲鴻
Xu Shiqing　徐世青
Xu Xiangqian　徐向前
xueming　學名

Ya'an　雅安
Yan Fu　嚴復
yang nigu　養尼姑
yanke duke biaoke　煙客
　　賭客婊客
yanqian　煙錢
Yao Yilin　姚依林
Yi (people)　彝
Yibin　宜賓
Yichang　宜昌
yi guo yuan li fei　以國原
　　利匪
yi shiye jiu guo　以實業救國
yishou　遺授
yixing　異姓
yi zhi wei zhu　以紙爲主
Yuanyi cooperative　遠益合
　　作社
yue　約

Yuechi county　岳池縣

yuhuo　預貨

zai shehui　在社會

zaofanpai　造反派

zaoshen tudi　灶神土地

zaozhi gongyequ　造紙工
　　業區

zaozhi shengchan hezuoshi
　　造紙生產合作社

zashui　雜稅

zerendi　責任地

zhai　宅

zhang　張

Zhang Daqian　張大千

Zhang Guotao　張國濤

Zhang Xianzhong　張獻忠

zhangbeizi　長輩子

Zhangyan village　張岩

Zhao Ziyang　趙紫陽

zhaogu　照顧

Zhengfu yao shanyu faxian
　　he huajie shehui maodun,
　　ba shezhong naoshi miao-
　　tou xiaochu zai mengya

zhuangtai　政府要善於
發現和化解社會矛盾,
把涉眾鬧事苗頭消除在
萌芽狀態

zhi fanzi　紙販子

zhipu　紙鋪

Zhiye tongye gonghui　紙業
　　同業公會

Zhong Chongmin　鏞崇敏

Zhonghua minzu de wenming
　　中華民族的文化

Zhongshansi　鐘山寺

Zhongxing　中興鄉

Zhu Guangrong　朱光榮

zhuma haozi　竹麻號子

Zigong　自貢

zigu weiwen fen you shui, er-
　　jin zhiyou pi wu juan　自
　　古未聞糞有稅, 而今只
　　有屁物捐

zili gengsheng　自力更生

ziliudi　自留地

Ziyang county　紫陽縣

zong citang　总祠堂

zong duobazi　总舵把子

B   Glossary of Selected Papermaking Terms

ENTRIES FOR PAPER TYPES, tools and equipment, raw materials, and miscellaneous terms are ordered alphabetically letter by letter, ignoring word and syllable breaks. Work processes are listed in the order they take place in production. For personal names and other terms, see Appendix A.

Paper Types

baozhi 報紙	newsprint
chuanlian 川連	Sichuan *lianshi*, small writing paper
dazhi 大紙	"big paper" of high quality, for writing and printing
duifang 對方	"matched squares," medium-quality writing paper
fanggu 仿古	"imitation antique" paper
gongbi 工筆	"meticulous brushwork," paper used for this style of painting
gongchuan 貢川	"tribute Sichuan," small writing paper
gongzhi 貢紙	"tribute paper," produced during the Qing for the imperial examinations
guohuazhi 國畫紙	"Chinese art paper," for brush painting and calligraphy
huangbiao 黃表	ritual paper produced in eastern Sichuan

jiaxuan 夾宣	"Jiajiang *xuan*," imitation *xuan* paper from Jiajiang
jingliao 精料	"pure pulp," high-quality *guohua* paper
lianshi 連史	large, fine writing paper
mingzhi 冥紙纸	"underworld paper," a generic term for ritual papers
mixinzhi 迷信紙	"superstition paper," for ritual purposes
pingsong 平松	a type of small, dyed ritual paper
shengxuan 生宣	"raw *xuan*," low-quality *guohua* paper
tulian 土連	type of ritual paper commonly produced in Hexi
tuzhi 土紙	"native paper," a general term for handmade paper
wenhuazhi 文化紙	"culture paper," writing and printing paper, esp. *baozhi* and *duifang* (qq.v.)
xiaozhi 小紙	"Small paper" of relatively poor quality, used as paper money and for other ritual purposes
xuanzhi 宣紙	*xuan* paper from Jingxian in Anhui, made from rice straw and tree bast
yangxiaoqing 洋小青	small ritual paper, dyed blue

Tools and Equipment

chizi, chijiao 池子, 池窖	soaking pool
duiwo 碓窩	pedal-operated hammer mill for pulping
huangguo 黃鍋鍋	traditional wooden bamboo steamer
lianchuang 簾床	rigid wooden frame to support the flexible molding screen

muzha 木榨	wooden press
shui bizi 水鼻子	"water nose," a strip of wood that wedges the molding screen to the frame
zhengguo 蒸锅鍋	modern pressure steamer made of steel-reinforced concrete
zhibi 紙纸壁	drying walls
zhicao 紙槽	molding vat
zhilian 紙知簾	paper mold or screen
zhi shuazi 紙刷子	paper brush used to paste the wet paper sheets onto the drying walls

Raw Materials

baijiazhu 白甲竹	bamboo (*Phyllostachys puberula makino*)
bianzhi 邊紙	paper cuttings, used as raw material
caojian 草鹼	potash, potassium carbonate (K_2CO_3)
chunjian 純鹼	soda ash, washing soda, sodium carbonate (Na_2CO_3)
huan tou 換頭	"changing the head," the process in which a bamboo plant flowers, withers, and grows back from side roots
huashui 滑水	mucilage, a plant extract that prevents the lumping of pulp fibers in the pulp-water solution
liaozi 料子	"stock" or "stuff," bamboo or *suocao* (q.v.) that has been reduced to soft fibers by soaking and steaming but has not yet been pulped
magen 马根馬根	the main rhizome of a bamboo plant

piaobaifen 漂白粉	chlorine bleach, sodium hypo-chlorite solution (NaClO) solution
shaojian 燒鹼	caustic soda, sodium hydroxide (NaOH)
shuizhu 水竹	bamboo (*Bambusa breviflora munro*)
suda 蘇打, *caoda* 曹達	baking soda, sodium bicarbonate (NaHCO$_3$)
suocao 蓑草, *longxucao* 龍鬚草	sabai grass (*Eulaliopsis binata*), now the main raw material for Jiajiang paper
zhijiang 紙漿	paper pulp
zhuma 竹麻	bamboo that is split, cut to size, soaked, and dried, used as raw material when no fresh bamboo is available

Miscellaneous

caozhi jiang 操紙匠	skilled paper molder (vatman)
dahu 大戶	big workshops employing wage labor and operating all year
peifang 配方	paper "recipe"
shuazhi jiang 刷紙匠	skilled paper brusher
xia cao 下槽	"reserving a vat": same as *yuhuo*
xianhuo 現貨	sales of paper or raw materials for cash (as opposed to *yuhuo*)
xiaohu 小戶	small workshops with little or no hired labor, usually seasonal
yuhuo 預貨	"bespoke goods": paper merchants advance cash or raw materials early in the season and are paid in paper at the end of the season

Work Processes

shui ou 水漚	soaking the split, cut bamboo in stone-lined ponds
zheng touguo 蒸頭鍋	first steaming: bamboo is soaked for a few more days in lime water or mixed with quicklime, then steamed for 6–7 days
da zhuma 打竹麻	after cooling for a day, the steamed bamboo in the steamer is pounded with large wooden pestles
xi liao 洗料	the steamed fibers are washed in ponds or mountain streams
zheng erguo 蒸二鍋	second steaming: the fibers are mixed with potash, soda ash, or caustic soda and steamed for 5 days, then rinsed several times to remove the soda
da bing 打餅	making "cakes": the steamed fibers are piled in tightly packed heaps and left to ferment for 20–30 days
zhi liao 制料 (pulp preparation)	
pulping 打漿	the *liaozi* is pulped in a tilt-hammer mill, hollander beater, or fodder cutter
rinsing 淘料	the pulp is washed several times in clear water; remaining dark fibers are removed
bleaching 漂白	pulp is mixed with chlorine bleach, left to stand for a day, then rinsed several times

cao zhi 操紙 (molding)

preparing emulsion 打槽	bleached pulp is added to the water in the vat and stirred vigorously
molding 操紙	paper sheets are formed with a mold, then rolled off ("couched") onto a pile of fresh sheets
pressing 壓水	at the end of the workday, the pile is put in the press and slowly squeezed dry

shua zhi 刷紙 (brushing)

preparing "folds" 打叠	the sheets are taken out of the press and separated with a pincer; ten sheets form a *die*, or "fold"
brushing 刷紙	wet sheets are brushed onto special drying walls; ten sheets form a *diao*
jiekai zhizhang 揭開紙張	when dry (after 1–2 days), the *diao* are taken down and the sheets are separated
finishing 整紙	the sheets are smoothed, checked for imperfections, cut to size, and packed for transport

C Main Paper Types and Their Markets in the Twentieth Century

Paper type	Size (cm)	Use	Main production area	Main market
I. LATE QING AND EARLY REPUBLIC (UNTIL CA. 1920)				
("Big paper": includes writing and printing paper [*lianshi*, *gongchuan*] and large-format dyed paper)				
Lianshi	122 × 68	Fine writing	Hedong	Chengdu and other cities in Sichuan
Gongchuan (tribute Sichuan)	60 × 25	Fine writing and ink painting	Hedong	Chengdu, Chongqing, Yunnan
Chuanlian (Sichuan *lianshi*)	49 × 22	Writing, printing	Hedong	Chengdu, Yunnan
Shuizhi (water paper)	73 × 45	Writing, printing	Hedong	All Sichuan
Hou/bo lanmei (thick/thin blue plum)	110 × 53, 95 × 51	Half-product. Dyed red, used for *jiashen* (ancestral scrolls)	Hedong	All Sichuan
Duiliao (matching pulp)	133 × 30	Half-product. Dyed red, used for *duilian* (matching couplets)	Hedong	All Sichuan

249

Paper type	Size (cm)	Use	Main production area	Main market

(*"Small paper"*: 40–50 types, used for paper money, name cards, wall paper, etc.; most common types)

Paper type	Size (cm)	Use	Main production area	Main market
Yinzhi (print paper)	33 × 18	Ritual paper money	Hexi	All Sichuan
Huang/bai zhonglian (yellow/white medium *lianshi*)	32 × 23	Ritual paper money	Hexi	All Sichuan
Huang tulian (yellow earth *lianshi*)	40 × 30	Ritual paper money	Hexi	All Sichuan

II. New types introduced after 1920*

Paper type	Size (cm)	Use	Main production area	Main market
Unbleached *duifang*	88 × 50	Writing, packing, fans, umbrellas, firecrackers	Spreads from Hedong to Hexi	All Sichuan
Bleached *duifang*	88 × 50	Writing, woodblock printing	Spreads from Hedong to Hexi	Chengdu, Chongqing
Baozhi (newsprint)	77 × 53	Machine printing	Spreads from Hedong to Hexi	Chengdu, Chongqing
Fangxuan (imitation *xuan*)	133 × 66, 166 × 83	Calligraphy, brush painting	Big work-shops in Hedong	Chengdu, Kunming

III. Main types of paper produced after 1950

Paper type	Use	Main market
Lianshe, gongchuan	Size and use: see above; phased out in the 1950s.	
Duifang (matched squares)	Size and use: see above. *Duifang* became the mainstay of the Jiajiang paper industry in the collective period.**	All China and export

Paper type	Size (cm)	Use	Main production area	Main market
Guohua (Chinese painting)†	100 × 53, 138 × 69, 153 × 84	Calligraphy, brush painting, decoration	Hedong	All China and export

IV. MAIN TYPES OF PAPER PRODUCED AFTER 1980

| *Duifang* (matched squares) | In Hedong, *duifang* was replaced by *guohua* paper in the 1980s. In Hexi, *duifang* production continued until ca. 1997, albeit at a small scale. | | | |

Guohua ("Chinese painting")

| *Shengxuan* | 100 × 53, 138 × 69, 153 × 84 | Calligraphy practice, decoration, gift wrapping | Hedong | All China and export |
| *Jingliao* | 180 × 97, 248 × 129 | Calligraphy and brush painting | Hedong | All China and export |

NOTES: *Baozhi* (newsprint) and *duifang* were collectively known as "culture paper" (*wenhua zhi*), because of their modern uses in printing and writing, in contrast to small-format "superstition paper" (*mixin zhi*). From 1920 on, they gradually replaced the older types of "big" writing paper (*lianshi, gongchuan*). In the 1930s, "small paper" producers in Hexi began to switch to unbleached *duifang* and newsprint.
**Duifang* became the mainstay of the Jiajiang paper industry in the collective area. Initially a medium-quality paper used for writing, wrapping, and other purposes, *duifang* acquired a new market as ritual paper after the different types of "small paper" had been banned. By the 1980s, *duifang* was used for sacrifices, firecrackers, and as blotting paper.
†Technically, *guohua* paper is a type of *shengxuan*, meaning that it is unsized (*sheng*, "raw") and resembles *xuan* paper from Anhui. Formats are the same as for Anhui *xuan* paper: 3 to 8 *chi*; 4 *chi* (138 × 69) is the most common size. It was developed by Shi Guoliang and Zhang Daqian in 1941 on the basis of *lianshi* paper, similar in format to 4-*chi xuan* paper. The name *guohua* was officially adopted in 1957. In the 1980s, producers of high-quality paper adopted the term *jingliao* ("pure pulp") to differentiate their paper, made from bamboo or grass fibers only, from poor quality *shengxuan* that contains large admixtures of waste paper.
SOURCES: Liang Binwen, "Sichuan zhiye diaocha baogao," 21–24; Su Shiliang, "Jiajiang xian zhiye zhi gaikuang," 7–9; Zhong Chongmin et al., *Sichuan shougong zhiye*, 25–28; Cheng Quan, "Jiajiang zhishi," 23–26.

Reference Matter

Notes

For complete author names, titles, and publication information on publications, see the Works Cited, pp. 283–313.

Introduction

1. See, among others, Fei Hsiao-t'ung, *Peasant Life in China*; Fei Hsiao-t'ung and Chang Chih-i, *Earthbound China*; Martin Yang, *A Chinese Village*; Fried, *Fabric of Chinese Society*; Anita Chan et al., *Chen Village Under Mao and Deng*; Edward Friedman et al., *Chinese Village, Socialist State*; idem, *Revolution, Resistance and Reform in Village China*; and Ruf, *Cadres and Kin*. Of these authors, Fei and Chang pay most attention to work. Cooper, *Artisans and Entrepreneurs of Dongyang*, is the only detailed English-language study of artisans in the PRC.

2. The literature on the rural-urban divide is huge and growing. See, among others, Chan Kam-Wing, *Cities with Invisible Walls*; Tiejun Cheng and Selden, "Origins and Social Consequences"; Knight and Song, *Rural-Urban Divide*; Potter, "The Position of Peasants"; Pun, "Becoming *Dagong-mei*"; Solinger, *Contesting Citizenship*; and Zhang Li, *Strangers in the City*.

3. Pomeranz, *The Great Divergence*; idem, "Women's Work, Family, and Economic Development in Europe and East Asia"; Li Bozhong, *Jiangnan de zaoqi gongyehua*; R. Bin Wong, *China Transformed*.

4. Feuerwerker, "Economic Trends, 1912–1949," 52.

5. Liu Ta-Chung and Kung-Chia Yeh, *Economy of the Chinese Mainland*, 66; Eastman, *Fields, Families, and Ancestors*, 87. Until 1935, China was largely shielded from the Great Depression by its adherence to the silver standard.

6. Liu Ta-Chung and Kung-Chia Yeh, *Economy of the Chinese Mainland*, 66.

7. Schwartz, *In Search of Wealth and Power*, 122–29; Bailey, *Strengthen the Country*, introduction.

255

8. "Economy" (*jingji*), "industry" (*gongye*), and "agriculture" (*nongye*) are loanwords or reintroductions from Japan. *Gongye* first appeared in the modern sense in 1877 and was used by Liang Qichao in 1896; *jingji* acquired its modern sense around 1901 (see Lydia Liu, *Translingual Practice*, 284, 290; and Masini, *Formation of Modern Chinese Lexicon*, 174).

9. James Scott, *Seeing Like a State*.

10. Kirby, "Engineering China," 152–53. See also Bian, *Making of the State Enterprise System*.

11. Liu Hongkang, *Zhongguo renkou: Sichuan fence*, 364. The difference for women was five cm.

12. Solinger, *Contesting Citizenship*.

13. Wang Fei-ling, "Reformed Migration Control and New Targeted People," 119–24; and Chan Kam Wing and Will Buckingham, "Is China Abolishing the *Hukou* System?" 604–6.

14. On *suzhi*, see Anagnost, "Corporeal Politics of Quality"; and Yan Hairong, "Neoliberal Govermentality." On hierarchies of value, see Herzfeld, *Body Impolitic*, chaps. 1 and 8.

15. On "peasantness" and "peasant consciousness" in contemporary China, see Flower, "Peasant Consciousness"; and Kipnis, "Within and Against Peasantness."

16. Cohen, "Cultural and Political Inventions," 154–55.

17. Mao Dun's short story "Spring Silkworms" is a good example of this view.

18. Zhang Ning, *Tudi de huanghun*, 15.

19. Zhang (ibid.) approvingly quotes Oswald Spengler's statement in *Decline of the West* that in peasant societies "man becomes plant again."

20. Sigaut, "Technology," 448.

21. See, e.g., Freedman, *Lineage Organization*; idem, *Chinese Lineage and Society*; and Faure, *The Structure of Chinese Rural Society*.

22. On "practical kinship," see Bourdieu, *Outline of a Theory of Practice*, 33–71.

23. Aristotle, *Politics*, II, 23.

24. Mann, "Household Handicrafts and State Policy," 77–79; Rowe, *Saving the World*, chap. 6.

25. Zelin, *Merchants of Zigong*, 331–32n20.

26. On goods and signs, see Appadurai, "Introduction"; Douglas and Isherwood, *World of Goods*; and Brewer and Porter, *Consumption and the World of Goods*.

27. Harrison, "Village Industries," 31–37.

28. Bian, *Making of the State Enterprise System*, 217–20; Kirby, "Engineering China," 152–53; Frazier, *Making of the Chinese Industrial Workplace*.

29. Knight and Song, *Rural-Urban Divide*, 27–34. Urban-rural income inequality has been rising throughout the reform period and was estimated at 3.36 : 1 in 2008.

30. Burawoy and Lukasz (*Radiant Past*, chap. 5) have argued that labor and its products were extracted in a much more visible way under state socialism than under capitalism, and that the very visible hand of the state produced a strong "negative class consciousness" among Eastern European workers. In a similar way, the rigidity of state planning gave credibility to the old labor slogan that "if your mighty arm so wills, all the wheels and gears stand still."

31. Marx, *Capital*, 1: 408, 451. See also Landes, "What Do Bosses Really Do," 588–93.

32. Braverman, *Labor and Monopoly Capital*.

33. For a discussion of skill among factory workers, see More, *Skill and the English Working Class*, 15–26. For skill in preindustrial work, see Sonenscher, *Hatters of Eighteenth-Century France*, 35–36.

34. Cockburn, *Brothers: Male Dominance and Technological Change*; Hafter, "Women Who Wove," 44; Tessa Liu, *Weaver's Knot*, 40–43, 234–38; Honeyman and Goodman, "Women's Work, Gender Conflict, and Labour Markets," 362–66; Joan W. Scott, "L'ouvrière: mot impie et sordide," 117–24.

35. Edwards, *Contested Terrain*; Andrew Friedman, *Industry and Labour*; Burawoy, *Manufacturing Consent*.

36. Sabel, *Work and Politics*.

37. Sigaut, "Technology," 445.

38. Koepp, "Alphabetical Order," 257.

39. Maier, "Between Taylorism and Technocracy"; Rabinbach, *The Human Motor*, 260–71; James Scott, *Seeing Like a State*, 97–102.

40. Bailes, "Alexei Gastev"; Rogger, "Amerikanizm and the Economic Development of Russia."

41. Donham, *Marxist Modern*, chap. 1.

42. Wright, "Spiritual Heritage of Chinese Capitalism"; Frazier, *Making of the Chinese Industrial Workplace*; Morgan, "Scientific Management in China."

43. This is Landes's ("What Do Bosses Really Do," 620) concise formula for the motive that drives technological change.

44. Kirby, "Engineering China," 152.

45. I rely for my understanding of Heidegger on Dreyfus, *Being-in-the-World*. See also Clark, *Being There*, 171.

46. I rely here on Andy Clark's overview in *Being There* of cognitive-science research over the past twenty years.

47. Ibid., 82.

48. Ibid., 191.

49. Ibid., chap. 3.

50. Hutchins, "Learning to Navigate," 62.

51. Lave and Wenger, *Situated Learning*, chap. 4.

52. Ingold, *Perception of the Environment*, 353.

53. Bourdieu, *Logic of Practice*, 57.

54. Bourdieu, *Outline of a Theory of Practice*, 78–90. A revised and shortened version of the argument can be found in *Logic of Practice*, chap. 3.

55. Bourdieu, *Logic of Practice*, 56. For a discussion of how the phenomenology of skilled factory work determined the response of German workers to the Nazi regime, see Lüdtke, "What Happened to the 'Fiery Red Glow'?"

56. Bourdieu, *Algérie 60*; Herzfeld, *Body Impolitic*; Willis, *Learning to Labour*.

57. The names of all interviewees have been altered to ensure their anonymity.

58. Sheng Yi and Yuan Dingji, "Jiajiang zaozhi." I accept the government's wish for secrecy and have omitted technical details, which, in any case, would be of little interest to most readers. Those interested in the technical aspects of the industry are referred to Pan Jixing, *Zhongguo zaozhi jishu shigao*, which contains a chapter on Jiajiang; and to Sheng Yi and Yuan Dingji, "Jiajiang zaozhi."

59. The point that the fieldwork situation is "a learning situation, in which the narrator has information which we lack" is forcefully made in Portelli, *Death of Luigi Trastulli*.

Chapter 1

1. Silkworm eggs are sold in "sheets" (*zhang*) of 100 eggs.

2. See Appendix B for a schematic overview of work processes in traditional papermaking and a list of technical terms. For more detailed descriptions of paper technology in Jiajiang, see Sheng Yi and Yuan Dingji, "Jiajiang zaozhi"; and Pan Jixing, *Zhongguo zaozhi jishu shigao*.

3. Western papermakers never learned how to prevent wet sheets from sticking together and therefore had to insert felt between the sheets.

4. In contrast to papermakers elsewhere in China, Jiajiang workshops did not use heated drying walls.

5. For the 115 workdays estimate, see Sichuan Archives, Gongyeting 1951 [13], 7. Zhong Chongmin et al., *Sichuan shougong zhiye*, estimates 138 days of vat work for 10,000 sheets of high-quality paper. No data are given for labor demand in steaming and finishing.

6. Notes 28/04/96; Interview Shi Haibo 11/04/96; Wang Shugong, *Jiajiang xianzhi*, 637. See also Shiga, "Family Property and the Law of Inheritance."

7. Interview Shi Xiujie and Zhang Wenshu 25/09/95.

8. Sheng Yi and Yuan Dingji, "Jiajiang zaozhi," 37.

9. Wolf and Huang, *Marriage and Adoption in China*, chaps. 5–7.

10. Wang Shugong, *Jiajiang xianzhi*, 636–37. The practice was also common in Dingxian, Hebei, where 70 percent of wives were older than their husbands (Gamble, *Ting Hsien*, 45).

11. Huang Fuyuan, *Zhima haozi*; Wang Shugong, *Jiajiang xianzhi*, 636.

12. Mendels, "Proto-Industrialization," 241–61.

13. Medick, "Village Spinning Bees," 319–25.

14. Huang Fuyuan, *Zhima haozi*.

15. Greek leather workers and other artisans, like Jiajiang papermakers, secretly approve of stubborn, recalcitrant, and "crafty" apprentices, even if they punish them for their acts. See Herzfeld, *Body Impolitic*, chaps. 2–5.

16. Interviews Shi Lanting 13/05/96, 17/09/98. For a discussion of space and gender in the Chinese house, see Bray, *Technology and Gender*, 51–83, 170–72.

17. Interview Shi Shengliang 21/10/95; Sheng Yi and Yuan Dingji, "Jiajiang zaozhi," 29.

18. Wang Shugong, *Jiajiang xianzhi*, 652.

19. Su Shiliang, "Jiajiang xian zhiye zhi gaikuang," 15.

20. Interviews Shi Dingliang 18/09/98, Shi Lanting 17/09/98, Shi Rongqing 19/04/96, and Shi Shenghuai 12/04/96.

21. *SCJJYK* 1936 [5: 2–3]. "Jiajiang gaikuang," 114. The same source gives the silver-copper exchange rate as 15,000–16,000 copper coins for one standard silver *yuan*. The unusually high exchange rate is probably due to the introduction of the *fabi* paper currency and the banning of private ownership of silver in late 1935, which drove up the price of silver coins.

22. Interview Shi Shenghuai 12/04/96.

23. Philip Huang, *Peasant Economy and Social Change in North China*, 310.

24. Interview Shi Dingliang 18/09/98. High marriage rates were also due to the low ratio of men to women in Jiajiang (93.6 to 100 in 1949), probably because of wartime conscription.

25. Sichuan Archives, Gongyeting 1951 [171: 1], 128–30.

26. Sharing of steamers is also reported by Fei Hsiao-t'ung and Chang Chih-i, *Earthbound China*, 177–96.

27. Sheng Yi and Yuan Dingji, "Jiajiang zaozhi," 20.

28. Interview Xie Changfu 22/09/98.

29. A good example for this comes from another papermaking area in China: in Jingxian, Anhui, two descent groups, the Caos and Wangs, more or less monopolized the production of *xuan* paper for 500 years (see Cao Tiansheng, *Zhongguo xuanzhi*, 43–56, 88–100).

30. Sheng Yi and Yuan Dingji, "Jiajiang zaozhi."

31. Ingold, *Perception of the Environment*, 353–54.

Chapter 2

1. Liu Zuoming, *Jiajiang xianzhi*, 251.

2. Wang Shugong, *Jiajiang xianzhi*, 58.

3. *Macun xiang dishibao diaochapu*. Wang Di ("Qingdai Sichuan renkou," 82) estimates that in Qing-period Sichuan, four *mu* were needed to feed one person.

4. *Jiajiang xian xiangzhen gaikuang*, 166–67. See Hosie, *Szechwan*; and Hsiang, "Mountain Economy," for a discussion of industries and sidelines in Sichuan mountain districts.

5. Liu Zuoming, *Jiajiang xianzhi*, 249; Wang Shugong, *Jiajiang xianzhi*, 2.

6. Gan Duan, *Jiajiang xian xiangtu zhilüe*, 2.10–17.

7. Lu Zijian, *Qingdai Sichuan caizheng shiliao*, 1: 742. For discussions of the demographic history of Sichuan, see Skinner, "Sichuan's Population in the Nineteenth Century"; and Li Shiping, *Sichuan renkoushi*, 146–79.

8. Xie Changfu, "Jiajiang xian Huatou xiangzhi," 62, 133, 169–76; Wang Shugong, *Jiajiang xianzhi*, 634.

9. Faure, *Structure of Chinese Rural Society*.

10. Li Shiping, *Sichuan renkoushi*, 136, 156–58; Entenmann, "Sichuan and Qing Migration Policy," 35–54. As William Rowe has shown, Macheng was an extremely violent place from the fourteenth to the twentieth century. In addition to breeding conflicts that spilled over into neighboring provinces, the county repeatedly served as a staging ground for rebel armies. For Macheng migration to Sichuan, see Rowe, *Crimson Rain*, 59–60, 64, 141, 151, 230.

11. Jiadangqiao stele, Shiyan village. See Chapter 9 for a fuller discussion of the stele.

12. A late Qing stele in Mianzhupu gives the woman's name as "née Hua" and dates her arrival to the Hongwu period (1368–98). Hongwu and

Wanli are well-known emperors, and their names may simply serve to underscore the claim of early, pre-Qing arrival.

13. Averill, "Shed People," 104–8.

14. Xie Changfu, "Jiajiang xian Huatou xiangzhi," 4. The text echoes the famous "Peach Garden Oath" from the *Romance of the Three Kingdoms*.

15. For discussions of *pengmin* settlers in Sichuan, see Vermeer, "The Mountain Frontier in Late Imperial China," 306–35; and Jerome Ch'en, *The Highlanders of Central China*. On *pengmin* elsewhere in China, see Rawski, "Agricultural Development"; Giersch, "A Motley Throng"; Averill, "Shed People"; Meskill, *A Chinese Pioneer Family*; and Pasternak, *Kinship and Community*.

16. Ruf, *Cadres and Kin*, chaps. 3–5.

17. Two characters, *yàn* 堰 and *nìan* 埝, were used interchangeably for the village name. Both mean "weir" or "embankment" and are pronounced *yàn* in the local dialect.

18. *Macun xiang dishibao diaochapu.*

19. Eastman, *Seeds of Destruction*, 149.

20. Interview Xiong Yuqing 21/10/95.

21. Wang Shugong, *Jiajiang xianzhi*, 643–44; Stapleton, "Urban Politics in an Age of 'Secret Societies'"; Skinner, "Aftermath of Communist Liberation," 65.

22. These were: literati (*ren*); wealthy merchants and landlords (*yi*); the middle classes (*li*); monks, priests, and fortunetellers (*zhi*); and butchers, hairdressers, and repairmen (*xin*) (see Zhang Wenhua, "Jiajiang xian Macun xiangzhi," p. 116).

23. Interviews Shi Dingliang and Zhang Wenhua 15–16/09/98, 18/09/98.

24. Interviews Shi Dingliang and Zhang Wenhua 15/09/98, 18/09/98.

25. For these associations, see Chapter 3.

26. *Jiajiang xian xiangtuzhi*, 24–38. See also Duara, *Culture, Power, and the State*, 148–57.

27. Graham, *Folk Religion in Southwest China*, 211–12.

28. Interview Xie Changfu 22/09/98.

29. Ruf, *Cadres and Kin*, 20; Naquin, *Peking: Temples and City Life*, chap. 7.

30. Xie Changfu, "Jiajiang xian Huatou xiangzhi," 172, 252–53; interview Xie Changfu 22/09/98.

31. Duara (*Culture, Power, and the State*, 122, 282) found that supravillage associations usually centered on market towns and included the villages of a marketing area. The Dongyue temple fair attracted people from a

wider area, but probably most of them were in some way linked to the market of Huatou.

32. Interview Xie Changfu 22/09/98.

33. Xie Changfu, "Jiajiang xian Huatou xiangzhi," 63, 253–56.

34. Yang Bingwen, "Zhuming da caohu Shi Ziqing," 26–27.

35. Xie Changfu, "Jiajiang xian Huatou xiangzhi," 68.

36. Zhang Wenhua, "Jiajiang xian Macun xiangzhi," 136–38; Wang Shugong, *Jiajiang xianzhi*, 703–10.

37. Freedman, *Lineage Organization*; idem, *Chinese Lineage and Society*. For a detailed review of the literature on Chinese kinship, see Santos, "The Anthropology of Chinese Kinship." For critical commentaries on Freedman, see Chun, "The Lineage-Village Complex"; James Watson, "Chinese Kinship Reconsidered"; and Ebrey and Watson, "Introduction," 1–2.

38. Detailed discussions can be found in the older literature on Chinese kinship: Feng Han-Yi, "The Chinese Kinship System"; Hu Hsien-Chin, *The Common Descent Group in China*; Chao Yuan Ren, "Chinese Terms of Address"; and Hui-chen Wang Liu, *The Traditional Chinese Clan Rules*. Most of these studies viewed kinship terminologies as classificatory systems. The rejection of this approach by fieldworking anthropologists has contributed to the neglect of *beifen* in Western studies since the 1950s.

39. My reading of Freedman here is influenced by Chun, "The Lineage-Village Complex," 430.

40. Rubie Watson, *Inequality Between Brothers*, chaps. 6, 7, and 10.

41. James Watson, "Chinese Kinship Reconsidered," 602.

42. To avoid terminological confusion, I follow Ebrey and Watson's ("Introduction," 5) suggestion to use "descent group" for groups characterized by demonstrated descent and ritual unity and to reserve "lineage" for those "descent groups that have a strong corporate basis in shared assets, usually, but not exclusively, land." Neither "lineage" nor "descent group" is ideal because both overemphasize lineal descent.

43. Wang Shugong, *Jiajiang xianzhi*, 634.

44. Ruf, *Cadres and Kin*, 51–53; Cohen, "Lineage Organization in North China," 521–28.

45. Interview Shi Dingliang 18/09/1998; Xie Changfu, "Jiajiang xian Huatou xiangzhi," 170–71, 192–93.

46. Interviews Shi Dongzhu 15/04/96; Shi Dingliang 28/04/96, 18/09/98; Shi Dingsheng 17/11/95.

47. Ruf, *Cadres and Kin*, 53.

48. The hall was razed during the Cultural Revolution.

49. In an argument between a Feng and a Wang, one of the opponents used the expression "fuck your mother" (*ri ni ma*). He received a beating and was fined 100 *jin* of lamp oil, not for foul language but for insulting a woman of his own family (interview Feng Youcai 26/09/98).

50. Ruf, *Cadres and Kin*, 173n29.

51. Xie Changfu, "Jiajiang xian Huatou xiangzhi," 245; Zhang Wenhua, "Jiajiang xian Macun xiangzhi," 120–21.

52. Interview Shi Dongzhu 15/04/96.

53. On gendered naming practices, see Rubie Watson, "The Named and the Nameless." Generational naming has been common throughout China since the sixth century (Ebrey, "Early Stages," 45–47; Davis, "Political Success," 86, 89–90).

54. To ensure the anonymity of interviewees, I have changed these generational markers.

55. See Davis, "Political Success," 89–90, for a discussion of generational naming in the Song dynasty. In his case, too, generational naming functioned as an "adhesive" that held the kinship group together; when the group declined and lost its corporate identity, generational names were abandoned.

56. It took me months to find out that two men whom I met almost daily (and who, as I learned later, shared a house and got along well) were brothers. Neither they nor anybody else drew attention to their close relationship, although their behavior clearly indicated that they belonged to the same generation.

57. Interview Shi Haibo 11/04/96; fieldnotes 28/04/96. For adoption and "dry kinship" (*gan qin*, a practice similar to godparentage, in which a "dry parent" gives a ritual name to a child, who thereby comes under his protection), see Ruf, *Cadres and Kin*, 35.

58. Baker, *Chinese Family and Kinship*, 15–16.

59. See Ruf, *Cadres and Kin*, 173n27.

60. Legend has it that on the day of Wanshun's wedding, his parents' house caught fire. Wanshun was accused of being a *baijiazi* (a son who has ruined his family), but when he sifted through the embers, he found a jar filled with silver from which he bought a new house and land in the plains. His grandsons, the "Seven Golds," kept seven silver ingots that were placed on the altar during the yearly sacrifices (interview Shi Shengyi 28/09/98).

61. This is most clearly expressed in the Jiadangqiao stele, discussed in Chapter 9.

62. Leach, *Pul Eliya*, 305.

Chapter 3

1. See Chapter 1 for estimates of labor productivity.

2. Fei Hsiao-t'ung and Chang Chih-i, *Earthbound China*, 177–96.

3. Interview Shi Guichun 11/04/96.

4. Interview Shi Dinggao 15/04/96.

5. This was expressed in a variety of ways: "thrice poor, thrice rich—neither lasts long" (*sanqiong sanfu bu dao lao*, perhaps a variation of the more common *sanqiong sanfu huo dao lao*, "thrice poor, thrice rich—one will live long"), or the enigmatic "thirty years east of the river, thirty years west of the river" (*sanshi nian hedong, sanshi nian hexi*).

6. Philip Huang, *Peasant Family and Rural Development*, 85, 111.

7. One male papermaker in Hexi described female wage workers as "old women of thirty or forty" (*san-sishi nian de laopo*). In this part of Sichuan, young women had their feet bound until the early 1930s, but not so severely that they could not stand or walk for long periods (interviews Shi Lanting 17/09/98, Xie Changfu 22/09/98).

8. Interviews Shi Haibo 11/04/96, 24/04/96; Shi Lanting 17/09/98; Shi Dingliang 19/09/98.

9. Interview Shi Rongqing 19/04/96.

10. Some big producers reportedly sold twice as much paper as they produced. It is unlikely that paper merchants were unaware of this practice, but since quality was vouchsafed by the seller, they did not object. The price for identical paper could differ by as much as 30 percent, depending on personal connections between buyer and seller (interviews Shi Haibo 11/04/96, 28/04/96; Shi Guichun 11/04/96).

11. Wang Shugong, *Jiajiang xianzhi*, 688–89; Zhang Wenhua, "Jiajiang xian Macun xiangzhi," 145–48; Yang Bingwen, "Zhuming da caohu Shi Ziqing."

12. European papermakers created watermarks by soldering a wire pattern on the rigid paper mold, which left an imprint on the sheet. Since Chinese molds are flexible, the pattern must also be flexible. Ziqing solved this problem by having a cloud pattern and his name embroidered on the mold with tiny glass pearls.

13. Both sentences echo the rhetoric of the Rights Recovery Movement, which opposed foreign control over Chinese resources (see Gerth, *China Made*, chap. 1). They also echo a saying by the Song-period general and patriot Yue Fei (1103–141): "Give us back our mountains and streams" (*huan wo he shan*).

14. *Macun xiang dishibao diaochapu*. Many households in Shiyan had substantial holdings outside the area, which do not appear in the survey. Shi Ziqing, for example, owned enough land to feed the 80–100 people permanently employed in his workshop and operated his own rice mill (interview Zhang Wenhua 16/09/98).

15. Xie Changfu, "Jiajiang xian Huatou xiangzhi," 88; interview Xie Changfu 22/09/98.

16. *Jiajiang xian xiangtuzhi*, 54.

17. Zhong Chongmin et al., *Sichuan shougong zhiye*, 37.

18. Su Shiliang, "Jiajiang xian zhiye zhi gaikuang," 7–8; Liang Binwen, "Sichuan zhiye diaocha baogao," 23–24.

19. Ren Zhijun, "Yishu Jiajiang," 1–8; Wang Shugong, *Jiajiang xianzhi*, 197, 225, 590; Wang Gang, "Qingdai Sichuan de zaozhi," 698–706.

20. Little, *Mount Omi and Beyond*, 223.

21. Wang Shugong, *Jiajiang xianzhi*, 236; Liang Binwen, "Sichuan zhiye diaocha baogao," 27. Modern maps show the Min River as a tributary and the Jinsha River as the main stream of the Yangzi, but the Sichuanese traditionally thought of the Min as the main stream.

22. Zhai Shiyuan, "Wo suo zhidao de Jiajiang zhi," 1.

23. Wang Shaoquan, *Sichuan neihe hangyun shi*, 230–32, 374–80.

24. Wang Lixian, *Sichuan gonglu jiaotong shi*, 49–90.

25. Ibid., 93–112.

26. Wang Shugong, *Jiajiang xianzhi*, 233.

27. Cheng Quan, "Jiajiang zhishi," 50.

28. Sichuan Archives, Jiansheting 1937 [1353a: 6]; Hua Younian, "Jiajiang zhi zhiye yu jinrong," 416; Zhong Chongmin et al., *Sichuan shougong zhiye*, 34–35.

29. Interview Ren Zhijun 09/05/96.

30. Zhai Shiyuan, "Wo suo zhidao de Jiajiang zhi"; interview Ren Zhijun and Xu Shiqing 23/09/98.

31. Interview Ren Zhijun 09/05/96.

32. Zhong Chongmin et al., *Sichuan shougong zhiye*, 34–35; Cheng Quan, "Jiajiang zhishi," 48–49.

33. Interview Ren Zhijun and Xu Shiqing 23/09/98.

34. Interviews Shi Haibo 11/04/96, Shi Guichun 11/04/96, Light Industry Bureau 07/05/96, Ren Zhijun 09/05/96, Shi Yongfan 13/05/96, Xie Changfu 21/09/98. Xie Changfu, the son of a Hexi papermaker, recalled that no interest was charged for loans repaid in less than one month. If the duration exceeded one month, 5 percent, and in a few cases up to 20 percent, was deducted. Zhong Chongmin et al. (*Sichuan shougong zhiye*, 48)

claim that as much as 50 percent of the current price was deducted, but this was strongly denied by my informants.

35. Interview Xie Changfu 21/09/98.

36. Zhong Chongmin et al., *Sichuan shougong zhiye*, 43–45; interview Shi Haibo 24/04/96.

37. Zhong Chongmin et al., *Sichuan shougong zhiye*, 48.

38. Ibid., 43–45, 54.

39. Hua Younian, "Jiajiang zhi zhiye yu jinrong"; *SCJJJK* 1944 [1: 3] "Yi-nianlai Chuansheng mijia."

40. Skinner, "Marketing and Social Structure" pt. I, 17–43; idem, "Cities and the Hierarchy of Local Systems," 275–88.

41. Skinner, "Cities and the Hierarchy of Local Systems," 275.

42. Ibid., 277–78.

43. Xie Changfu, "Jiajiang xian Huatou xiangzhi," 4.

44. Zhang Wenhua, "Jiajiang xian Macun xiangzhi," 138.

45. Interview Ren Zhijun and Xu Shiqing 23/09/98.

46. Skinner, "Marketing and Social Structure," pt. I, 32.

47. Philip Huang, *Peasant Economy and Social Change in North China*, 220–22.

48. Skinner, "Marketing and Social Structure," pt. I, 35.

49. Interview Shi Dingliang 18/09/98. Roasted rice (*huomi*) was the staple food in the paper districts. Roasting reduced the weight of the rice and protected it from mold, insects, and rodents. Some people also preferred the taste.

50. Interview Xie Changfu 21/09/98.

51. Interview Shi Dingliang 18/09/98.

52. Interviews Shi Dingliang 18/09/98, 27/09/98.

53. Zhang Wenhua, "Jiajiang xian Macun xiangzhi," 20, 25–26, 102–3, 132–34. Chuanzhu is variously identified as Li Bing, the Qin dynasty official and builder of the Dujiangyan irrigation system; as his son Li Erlang; or as the Song dynasty official Zhao Yu—all of them upright officials who improved irrigated agriculture in Sichuan.

54. Zhang Wenhua, "Jiajiang xian Macun xiangzhi," 115–17.

55. Interview Shi Dingliang 18/09/98.

Chapter 4

1. One such charity is described in Rowe, *Hankow: Conflict and Community*, 103–4. See also Ts'ien, *Paper and Printing*, 109.

2. The "four great inventions" are the compass, gunpowder, paper, and printing.

3. Mazumdar, *Sugar and Society in China*, 211–17.

4. Mann, *Precious Records*, 246.

5. Lean, "One Part Cow Fat, Two Parts Soda," 7–11.

6. MacGowan, "Chinese Guilds," 182. Since the master was under protection of the magistrate, "the word was passed around that 'biting to death is no murder.' Gild members to the number of 120 each took a bite, no one being allowed to leave the place whose lips and teeth were not bloody." The story is repeated in much of the later literature, including Max Weber's *Religion of China*.

7. See, e.g., Liu Min, "Sichuan shehui jingji," 102–8; and Quan Hansheng, *Zhongguo hanghui zhidu*, 201–3, 205–10.

8. Wright, "Distant Thunder," 702–4; Bramall, *In Praise of Maoist Economic Planning*; Kapp, "Chungking as a Center of Warlord Power"; Zhang Xuejun and Zhang Lihong, *Sichuan jindai gongyeshi*, chap. 6.

9. Kapp, *Sichuan and the Chinese Republic*, 88–105.

10. Ibid., 93; Bramall, *In Praise of Maoist Economic Planning*, 282–91.

11. Kirby, "Engineering China"; idem, *Germany and Republican China*; Zanasi, *Saving the Nation*, chap. 1; Bian, *The Making of the State Enterprise System*, chap. 7.

12. The fact that the Jiajiang industry attracted the attention of the Qing state so early in the dynasty suggests that it dates back to the Ming, although there is no firm documentary evidence for this (see Li Yubing, "Jiajiang chanzhi ying shiyu heshi?").

13. Liu Zuoming, *Jiajiang xianzhi*, 31.

14. Ibid., 54–59, 110, 113; Xie Changfu, "Jiajiang xian Huatou xiangzhi," 176–78.

15. Yiwanshui stele, 1855; see also Xie Changfu, "Jiajiang xian Huatou xiangzhi," 61, 178.

16. Tu Song, *Jiajiang xianzhi*, *fengsu* section.

17. Liu Zuoming, *Jiajiang xianzhi*, 31.

18. *SCYB* 1936 [9: 1], "Jianting diaocha Jia-Hong-E," 122.

19. Interviews Zhang Wenhua 19/09/98, Light Industry Bureau 05/09/96; Liao Tailing 14/05/96.

20. Similar notions of skill as a communal property, vested in self-regulated corporations and accessible only to individuals qualified through birth or apprenticeship, also existed in Western Europe (see Rule, "The Property of Skill"; and Somers, "The 'Misteries' of Property").

21. Duara, *Culture, Power, and the State*, 25.

22. Ibid., 1–4, 58–61.

23. Ibid., 73–77.

24. *Jiajiang xian xiangtuzhi, dili* section; Liu Zuoming, *Jiajiang xianzhi*, 47.

25. Interview Xie Changfu 22/09/98; Xie Changfu, "Jiajiang xian Huatou xiangzhi," 67, 277.

26. Su Shiliang, "Jiajiang xian zhiye zhi gaikuang"; *SP* 1925, "Sichuan Jiajiang xian zhi zhiye," 40.

27. Sichuan Archives, Jiansheting 1937 [1353a: 6]; Liang Binwen, "Sichuan zhiye diaocha baogao," 29.

28. Zhong Chongmin et al., *Sichuan shougong zhiye*, 46–47.

29. Interviews Ren Zhijun 09/05/96, Ren Zhijun and Xu Shiqing 23/09/98, Liao Tailing 14/05/96.

30. Liu Shaoquan, *Jiajiang de zhiye yu guoji jiaoliu*, 29; Xie Changfu, "Jiajiang xian Huatou xiangzhi," 160.

31. Wang Shugong, *Jiajiang xianzhi*, 226.

32. Interview Ren Zhijun 09/05/96.

33. Interview Ren Zhijun and Xu Shiqing 23/09/98.

34. Wang Shugong, *Jiajiang xianzhi*, 420–21. On late Qing chambers of commerce, see Chen Zhongping, "The Origins of Chinese Chambers of Commerce"; and Fewsmith, "From Guild to Interest Group."

35. Interview Ren Zhijun 09/05/96.

36. Peng, who joined the underground CCP in 1931, continued his career after 1949. He became a leading member of the provincial association of industry and commerce and a member of the national congress. He died in 1968, at the age of 93, after Red Guard attacks on his family. Two of his sons, a niece, and a nephew joined the CCP in Yan'an (Wang Shugong, *Jiajiang xianzhi*, 679–81, 696–700).

37. Sichuan Archives, Jiansheting 1936 [1353a: 1]; Jiansheting 1939 [4042: 1, 2]; Li Yubing and Lei Yinglan, "Zhiye jushang Xie Rongchang," 1–9.

38. Sichuan Archives, Jiansheting 1937 [1353a: 6].

39. Sichuan Archives, Jiansheting 1937 [1353a: 7].

40. Sichuan Archives, Jiansheting 1937 [1353a: 12].

41. Eastman, *Seeds of Destruction*, 50.

42. Wang Shugong, *Jiajiang xianzhi*, 298; Xie Changfu, "Jiajiang xian Huatou xiangzhi," 160. Actual collection fell short of these rates.

43. Sichuan Archives, Jiansheting 1942 [9338: 1].

44. Sichuan Archives, Jiansheting 1942 [1353b: 1]. See also Hung, *War and Popular Culture*, 151–86, for a discussion of how newspapers were used for mobilization against Japan.

45. Wang Shugong, *Jiajiang xianzhi*, 298.

46. Albert Feuerwerker, "Economic Trends, 1912–1949," 55.

47. Most of the documents discussed in the following pages were published by the research institutes of the Bank of China (*Sichuan yuebao*), the Provincial Bank of Sichuan (*Sichuan jingji yuekan* and *Sichuan jingji jikan*), or by the Agricultural Bank (Zhong Chongmin's monograph on the paper industry) and aimed at an audience of government officials and functionaries in the banking sector. I found about fifty articles on papermaking from the years 1933 to 1947. In addition to the Jiajiang paper industry, they discuss papermaking in Liangshan, Tongliang, Guang'an, Hechuan, Dazhu, and Baxian counties.

48. Lü Pingdeng, *Sichuan nongcun jingji*, 334. Imports from Europe and the United States were replaced in the 1910s–1920s by cheap Japanese paper. After 1931, the boycott of Japanese goods put an end to most paper imports. The nascent modern paper industry of coastal China was too small to contribute much to imports.

49. Gan Cisen, *Zuijin 45 nianlai*.

50. Sichuan xinwen chubanju, Shizhi bianzou weiyuanhui, *Sichuan xinwen chuban shiliao*, 52, 129–31, 165. For a discussion of the printing industry of Yuechi, Sichuan's main printing center, see Brokaw, *Commerce in Culture*, 540–44.

51. Zhong Chongmin et al., *Sichuan shougong zhiye*, 49.

52. *SCJJYK* 1935 [3: 2] "Jiajiang zhiye diaocha," 89.

53. *SCJJYK* 1935 [3: 1] "Jiajiang gailiang zhiye"; Hua Younian, "Jiajiang de zhiye yu jinrong," 418–19; Liu Zidong, "Sanshisannian Jiajiang jingji dongtai," 201.

54. In Jiajiang, the customary amount of paper burned at funerals was 144,000 sheets, plus paper effigies, clothes, etc. Those who burned less "risked the contempt of relatives, neighbors, and friends" (Gan Duan, *Jiajiang xian xiangtu zhilüe*, 17).

55. Sichuan Archives, Jiansheting 1943 [9338: 5]. The text does not explain how the production of "superstition paper" benefited the Communists. In fact, Jiajiang no longer produced such paper in the 1940s.

56. Hua Younian, "Jiajiang de zhiye yu jinrong," 415–19.

57. Liu Zidong, "Sanshisannian Jiajiang jingji dongtai."

58. SCYB 1935 [6: 3], "Shengfu shixing zhenxing," 159–60; *SCJJYK* 1935 [3: 4–5], "Jianting shishi zhizhi"; *SCJJYK* 1935 [3: 2], "Jiajiang zhiye diaocha," 89.

59. The great stands of softwood timber in present-day Liangshan, Aba, and Ganzi prefectures were inaccessible before the construction of motor roads in the 1950s.

60. Li Jiwei, "Gailiang Jiajiang zaozhiye," 18–20; Duan Zhiyi, "Sichuan shougong zaozhiye," 39–40; Qiu Xian, "Zhenxing zaozhi gongye."

61. Jiang Huice, "Sichuan xi'nanqu jingji jianshe couyi," 93.

62. *SCJJYK* 1935 [3: 1], "Jiajiang gailiang zhiye."

63. Jiang Huice, "Sichuan xi'nanqu jingji jianshe couyi."

64. Alitto, *The Last Confucian*; Hayford, *To the People*; Fitzgerald, "Warlords, Bullies, and State Building"; Zanasi, "Exporting Development," 157–63; idem, *Saving the Nation*, 109–15, 133–73.

65. Fitzgerald, "Warlords, Bullies, and State Building," 431–32.

66. Sichuan Archives, Sichuan hezuo jinku 1941.

67. Interview Zhang Wenhua 16/09/98; Wang Shugong, *Jiajiang xianzhi*, 704–5; Sichuan Archives, Jiansheting 1945 [5117: 1].

68. Liu Min, "Sichuan shehui jingji," 102, 108.

69. Jiang Huice, "Sichuan xi'nanqu jingji jianshe couyi."

70. Liang Binwen, "Sichuan zhiye diaocha baogao"; Zhang Xiaomei, *Sichuan jingji cankao ziliao*, section R, 102–7.

71. *SCYB* 1934 [5: 6], "Jiajiang zhizhi gongye gaikuang."

72. Zhang Xuejun and Zhang Lihong, *Sichuan jindai gongyeshi*, 179.

73. *SCYB* 1932 [1: 3], "Gedi zaozhiye gaikuang."

74. *SCJJYK* 1935 [3: 2], "Jiajiang zhiye diaocha."

75. Su Shiliang, "Jiajiang xian zhiye zhi gaikuang," 8; Ren Zhijun, "Jiajiang shougong zhi de chanxiao gaikuang," 3–4.

Chapter 5

1. Sichuan Archives, Gongyeting 1951 [171: 1], 133.

2. These were heavily inflated "old" *yuan renminbi* (Wang Shugong, *Jiajiang xianzhi*, 220).

3. Veilleux, *Paper Industry in China*, 7–13.

4. Sichuan Archives, Gongyeting 1951 [19: 5].

5. Sichuan Archives, Gongyeting 1952 [106].

6. Sichuan Archives, Gongyeting 1951 [19: 1], 40.

7. Liu Ta-Chung and Kung-Chia Yeh, *Economy of the Chinese Mainland*, 88, 209; Emerson, *Non-Agricultural Employment*, 83, 128. In 1953, Zhu De ("Ba shougongyezhe zuzhi qilai," 100) cited a number of 19.3 million handicraft workers.

8. Zhonghua quanguo shougongye hezuo zongshe and Zhonggong, Zhongyang dangshi yanjiushi, *Zhongguo Shougongye hezuohua*, table 4, 1: 708.

9. Wang Haibo, *Xin Zhongguo gongye jingji shi*, 386. See also Eyferth, "Introduction," 7–11.

10. Schran, "Handicrafts in Communist China," 152–53.

11. Perry, "From Native Place to Workplace," 44–48.

12. Liu Shaoqi, "Guanyu xin Zhongguo de jingji jianshe fangzhen," 27. The original formulation is from Lenin, whom Liu quotes.

13. Liu Shaoqi, "Guanyu shougongye hezuoshe wenti," 105.

14. Zhonggong, Zhongyang, "Zhonggong Zhongyang guanyu xunsu huifu," 185.

15. The Southwest Administrative Region included Sichuan, Guizhou, Yunnan, and the short-lived province of Xikang. The West Sichuan region comprised most of the Chengdu basin. Both the supra-provincial regions and the subprovincial districts were abolished when the PRC returned to the provincial model in 1954 (see Solinger, *Regional Government*, chap. 1).

16. The administrative history of the paper industry is too tangled to be described in detail. Administrative oversight began with a "papermaking guidance committee" organized by the West Sichuan Industry Bureau, which later evolved into the county-level Handicraft Bureau, renamed in 1976 the Second Light Industry Bureau ("second light industry" refers to unmechanized and collective industries; mechanized, state-owned industries belonged to the first light industry system). The Handicraft/SLI Bureau had a dual affiliation with the supply-and-marketing cooperatives and the Light Industry administration. During the Great Leap Forward and the Cultural Revolution, the Handicraft/SLI Bureau was remerged into the Industry Bureau.

17. Sichuan Archives, Gongyeting 1951 [19: 4], 61.

18. Lardy, *Agriculture in China's Modern Economic Development*, 48–50.

19. Sichuan Archives, Gongyeting 1951 [13].

20. Ibid., 16–18.

21. Mao Zedong, "Speed up the Socialist Transformation of Handicrafts," 283.

22. Sichuan Archives, Gongyeting 1951 [93: 6].

23. Fiber yield in traditional papermaking was around 15 percent; i.e., 100 kg of fresh bamboo yielded 15 kg of fiber. Collective workshops aimed to increase that yield to 30 to 40 percent. This could be achieved through more aggressive chemicals (which lead to a more complete digestion of the fibers) and reduced washing of the pulp (washing leads to the loss of short fibers). The results are a higher acidity and a higher percentage of short and brittle fibers (Sichuan Archives, Gongyeting 1950 [19: 3], 47).

24. Ibid.

25. Solinger, *Regional Government*, 31–34; Skinner, "Aftermath of Communist Liberation," 61–76.

26. Sichuan Archives, Gongyeting 1951 [171: 1], 133; Wang Shugong, *Jiajiang xianzhi*, 429, 446. On anticommunist resistance in southwest China, see also Brown, "From Resisting Communists to Resisting America," 113–25.

27. Interview Shi Dongzhu 28/10/95.

28. Interviews Shi Dongzhu 28/10/95, Shi Shengfan 26/09/95, Shi Dingmin and Zhang Wenhua 15/09/98.

29. Xie Changfu, "Jiajiang xian Huatou xiangzhi," 70; interview Xie Changfu 21/09/98.

30. Interviews Shi Dingliang 18/09/98, Shi Dongzhu 24/04/96.

31. Mao Zedong, "How to Differentiate the Classes in the Rural Areas." In this text, Mao recognized the existence of workers (*gongren*) in the countryside. Later texts often substitute "agricultural laborers" (*gunong*) for workers.

32. Mao Zedong, *Report from Xunwu*, 99, 109–12.

33. Zhongyang renmin zhengfu, "Zhonghua renmin gonghehuo tudi gaige fa," 3, 5, 7; idem, "Zhongyang renmin zhengfu Zhengwuyuan," 37, 56–58. See also the English translation of land reform documents in Hinton, *Fanshen*, 727–40.

34. Xie Changfu, "Jiajiang xian Huatou xiangzhi," p. 74.

35. Kuhn, "Chinese Views of Social Stratification"; Billeter, "The System of 'Class Status.'"

36. Interview Shi Rongqing 19/04/96.

37. Interviews Shi Dongzhu 24/04/96, Xie Baoqing 02/05/96.

38. Interviews Shi Dongzhu 24/04/96, Shi Haibo 24/04/96. One way to expropriate the nominally protected property of industrialists was to make them pay debts and back wages they owed to workers.

39. For similar cases from elsewhere in Sichuan, cf. Ruf, *Cadres and Kin*, 86–87; and Endicott, *Red Earth*, 25–26.

40. Xie Changfu, "Jiajiang xian Huatou xiangzhi," 74–75, 278.

41. Interviews Shi Dinggao 03/11/95; Shi Dingmin 07/11/95; Shi Dongzhu 03/10/95, 13/11/95; Shi Lanting 08/11/95.

42. The old *bao* were renamed *cun* (village) in 1952 (interview Xie Baoqing 02/05/96).

43. Cheng Quan, "Jiajiang zhishi," 10.

44. Interview Xie Baoqing 02/05/96.

45. Interviews Shi Lanting 08/11/95; Shi Dingmin 7/11/95; Shi Dongzhu 03/10/95, 13/11/95.

46. Riskin, "Small Industry and the Chinese Model of Development," 263. See also idem, "China's Rural Industries," 78–83.

47. Interview Xie Baoqing 02/05/96.

48. Sichuan Archives, Gongyeting 1951 [19: 2], 20; Gongyeting 1951 [171: 1], 132. On the intricacies of rural-urban and agricultural-nonagricultural definitions, see L. Zhang and Simon Zhao, "Re-examining China's 'Urban' Concept"; and Martin, "Defining China's Rural Population."

49. Sichuan Archives, Gongyeting 1951 [13].

50. Interview Xie Baoqing 02/05/96.

51. Interview Zhang Xuelin 22/09/98.

52. Interview Shi Dingmin 07/11/96.

53. Like urban people, grain recipients in the paper districts had "grain booklets" (liangben), in which their rations and eventual overdrafts were noted.

54. Interview Shi Lanting 08/11/95.

55. Interview Shi Dingmin 16/10/95.

56. The Jiajiang model workshop established in 1951 remained operational until 1966, although lack of funds and poor management forced it to suspend operations several times.

57. See, e.g., Du Shihua, "Shougong zhujiang de zhizao"; Lu Deheng, "Bian shougongzhi shengchan"; idem, "Shougongzhi shengchan zhong"; and Sichuan sheng shougongzhi, "Shougongzhi shengchan zhong shiyong."

58. Interviews Wang Yulan 24/10/95, Shi Lanting 23/10/95.

59. A similar argument is often made with reference to Hexi workshops and the poorer quality of their paper: Hexi producers are incapable of fine, careful work because of their different "soil and water."

60. Mitchell, Rule of Experts; James Scott, Seeing Like a State.

61. Kotkin, Magnetic Mountain; Siegelbaum and Suny, Making Workers Soviet. For a Chinese case, see Rofel, Other Modernities.

62. Burawoy and Krotov, "Soviet Transition from Socialism to Capitalism."

Chapter 6

1. Kane, Famine in China; Banister, China's Changing Population, 59–60, 83, 85.

2. Bramall, In Praise of Maoist Economic Planning, 297. Cao Shuji ("1958–61 nian Sichuan renkou siwang," 1) calculates 9.4 million excess deaths, equivalent to 13 percent of the province's population.

3. Potter and Potter, China's Peasants, 68–82; Domenach, Origins of the Great Leap Forward; Yang Dali, Calamity and Reform in China, chap. 1.

4. Zhang Wenhua, "Jiajiang xian Macun xiangzhi," 38.

5. These stalls are among the few fond memories people have of the period. For a long time, people paid for everything they consumed, putting the appropriate sum in a box. When children stole a few biscuits, this caused a major scandal (interview Shi Rongxuan 02/11/95).

6. Wang Shugong, *Jiajiang xianzhi*, 503.

7. Ibid., 143.

8. Interview Shi Dongzhu 31/10/95.

9. Ibid.

10. Walker, *Food Grain Procurement*, 8–81, 94; Bramall, *In Praise of Maoist Planning*, 325.

11. Wang Shugong, *Jiajiang xianzhi*, 82.

12. Bramall, *In Praise of Maoist Planning*, 293.

13. Interview Shi Dinggao 07/11/95; Wang Shugong, *Jiajiang xianzhi*, 270, 567.

14. Interview Peng Chunbin 12/11/95.

15. *Jiajiang xian xiangzhen gaikuang* 1991. I had no access to year-by-year, township-level data. Population losses are calculated from an internal publication that gives demographic data for the 1950s (in most townships for 1953, sometimes earlier or later) and for the 1960s (mostly for 1964). For the data and their limitations, see Eyferth, "De-industrialization in the Chinese Countryside."

16. Macun, with a high percentage of papermakers and losses of only 8 percent, is an exception to this pattern. Proximity to the county seat, good connections to county officials, and Macun's status as an (unofficial) model for the paper industry may have contributed to its relatively low mortality.

17. The general rule was that no more than 7 percent of the total arable land should be privately used. In Macun, this would have resulted in private plots of a few square meters per capita.

18. Interviews Shi Dingliang 22/04/96, Shi Dinggao 07/11/95; Cheng Quan, "Jiajiang zhishi," 12.

19. Selden, *People's Republic of China*, 522–23.

20. Zhonggong, Zhongyang, "Zhongguo Zhongyang guanyu chengxiang shougongye," 247. This document was more widely known as the "Thirty-Five Articles on Handicrafts" ("Shougongye sanshiwu tiao").

21. Zhongyang shougong guanli zongju, "Guanyu 1963 nian jinyibu kaizhan zhengshe," 308.

22. Wang Shugong, *Jiajiang xianzhi*, 221.

23. Cheng Quan, "Jiajiang zhishi," 12.

24. Interviews Shi Dinggao 07/11/95, Zhang Xuelin 04/12/95.

25. Wang Shugong, *Jiajiang xianzhi*, 91.

26. Interviews Xie Baoqing 02/05/96, 06/05/96; Yang Yuanjin 22/09/98.

27. Cf. Anita Chan and Unger, "Grey and Black."

28. Cheng Quan, "Jiajiang zhishi," 56; interviews Xie Baoqing 02/05/96, 06/05/96.

29. Lardy, *Agriculture in China's Modern Economic Development*, 49–50.

30. Yang Dali, *Calamity and Reform in China*, 109–14.

31. Wang Shugong, *Jiajiang xianzhi*, 269.

32. Interviews Xie Baoqing 02/05/96, 06/05/96.

33. Wang Shugong, *Jiajiang xianzhi*, 124–25. This is higher than the average national procurement rates of 13–17 percent in the 1970s (see Oi, *State and Peasant*, 63).

34. Interview Light Industry Bureau 07/05/96.

35. Cheng Quan, "Jiajiang zhishi," 57–58; Wang Shugong, *Jiajiang xianzhi*, 356.

36. This quota fluctuated slightly in later years.

37. Interviews Shi Chunjin 29/11/95, Shi Shengyan 30/11/95.

38. Wang Shugong, *Jiajiang xianzhi*, 270; interview Shi Dinggao 16/11/95.

39. Zhonghua quanguo shougongye hezuo zongshe, "Quanguo shougongye hezuo zongshe guanyu zhengdun," 295; Sichuan Archives, Jianchuan 1963 [074-99b], 9–12; Jianchuan 1963 [074-99e], 3–5; Jianchuan 1963 [074-17], 4–8, 9–12.

40. Sichuan Archives, Jianchuan 1960 [80-3111], 1–12; Jianchuan 1963 [074-17].

41. Sichuan Archives, Jianchuan 1963 [074-17], 6, 7–9.

42. Veilleux, *Paper Industry in China*, 28, 46.

43. Ibid., 1.

44. Wang Shugong, *Jiajiang xianzhi*, 228.

45. Veilleux, *Paper Industry in China*, 71.

46. Ibid., 130.

47. Cheng Quan, "Jiajiang zhishi," 63; interviews Xie Baoqing 30/04/96, 06/05/96; Xiao Zhicheng 25/09/98.

48. Ma Mingzhang, "Luoshi quanshu, lin nong huanwei," 12.

49. Interview Shi Dinggao 16/11/95.

50. In Shandong in the 1940s, for example, "sweet potato eater" was considered an insult (see Martin Yang, *A Chinese Village*, 34–35).

51. Papermakers elsewhere in China experienced the loss of their trade as equally painful. In Chiqiao, Shanxi, former papermakers often asked to be buried with their tools, "as a mark of the respect they once had for a skill that is no longer needed" (see Harrison, *Man Awakened from Dreams*, 168).

52. Interview Shi Shengji 03/10/95. Shi Shengji inadvertently echoed Philip Huang's involution argument in *Peasant Family and Rural Development* (see p. 317).

53. The houses themselves were privately owned, although housing space and construction sites and materials were increasingly allocated by team leaders.

54. Interviews Shi Lanting 23/10/95, 13/05/96, 17/09/98; Xu Shimei 25/10/95.

55. Interviews Shi Shengyi 18/11/95, 28/09/98; Shi Shengliang 27/11/95.

56. Wang Shugong, *Jiajiang xianzhi*, 217.

57. *Jiajiang turang*, 133.

58. Sigurdson, *Rural Industrialization in China*; Riskin, "Small Industry and the Chinese Model of Development"; American Rural Small-Scale Industry Delegation, *Rural Small-Scale Industry*.

59. Dali Yang (*Calamity and Reform in China*, 109–14) describes this process as "backward specialization."

60. Wang Shugong, *Jiajiang xianzhi*, 98–100, 109–11, 124–25.

61. Ibid., 177–79, 181; *Jiajiang xian xiangzhen gaikuang*, 15–24, 55–88. See also Donnithorne, "Sichuan's Agriculture: Depression and Revival," 67–75.

62. Wang Shugong, *Jiajiang xianzhi*, 210–11.

Chapter 7

1. Cheng Quan, "Jiajiang zhishi," app. B.

2. Ibid., 57–58.

3. Wang Shugong, *Jiajiang xianzhi*, 196.

4. Ruf, *Cadres and Kin*; Endicott, *Red Earth*, 134–36.

5. Interview Shi Dinggao 04/12/95.

6. The years 1983–84 experienced easy credit, as control over bank loans was loosened to stimulate growth (see Yuan Peng, "Capital Formation in Rural Enterprises," 109).

7. Interview Shi Shengji 14/11/95.

8. Different kinds of waste paper differ considerably in quality. The best waste paper comes from print shops (cuttings of strong, white, unprinted paper) and engineering firms (cuttings of blueprint paper).

9. Zhang Wenhua, "Jiajiang xian Macun xiangzhi," 75–76.

10. Interview Shi Jianrong 01/12/95.

11. Interview Shi Lanting and Shi Shengliang 02/12/1995.

12. This is net household income as calculated by the papermakers: sales income minus material costs, steaming fees, wages, and wage-related costs.

Household labor costs, depreciation of investment, and interest payments on loans are not included as costs.

13. *Jiajiang xian 1994 nian nongye shengchan*.

14. The Xias' workshop described in Chapter 1 is exceptional in this respect.

15. Strictly speaking, *shengxuan* refers to all unsized, water-absorbent handmade paper, but in Jiajiang, the term is now used only for cheaper paper made mainly from waste paper. Such designations are often subject to "brand-name inflation," as cheap imitations appear and quality producers shift to new, more exclusive names.

16. Peng Xizhe, *Demographic Transition in China*, 223.

17. Wang Shugong, *Jiajiang xianzhi*, 81; Sichuan sheng Jiajiang xian bianshi xiuzhi weiyuanhui, *Jiajiang xian nianjian*, 1995, 88.

18. Interviews Shi Lanting 23/10/95, 09/11/95.

19. Interview Shi Yongchun 04/11/95.

20. See Kipnis, *Producing Guanxi*, 138.

21. See Shiga, "Family Property and the Law of Inheritance"; Cohen, *House United, House Divided*; and idem, *Kinship, Contract, Community, and State*, chap. 4. Due to the one-child policy (strictly implemented since 1980), families with more than one son have become rare. At the time of my interviews, however, many families still had two or more sons in their teens or twenties who were still living with their parents.

22. Judd, *Gender and Power in Rural North China*, 175.

23. Interview Xu Shimei 25/10/95.

24. Judd, *Gender and Power in Rural North China*, 182 (italics in original).

25. Ibid., 137–53.

26. Anagnost, *National Past-Times*, chap. 3. See Wang Xiaoqiang and Bai Nanfeng, *The Poverty of Plenty*, 23–65, 154–74, on the lack of "market consciousness" among peasants and minority populations.

Chapter 8

1. Wang Shaoguang, "The Politics of Private Time," 152–58.

2. Kraus, *Brushes with Power*, chaps. 8–9, 13–14.

3. Liu Shaoquan, *Jiajiang de zhiye yu guoji jiaoliu*, 36.

4. Cheng Quan, "Jiajiang zhishi," 56.

5. Ibid., 55, 59.

6. This was part of a larger and generously funded research project on the history of Chinese papermaking.

7. Zhang Wanshu, "Guanyu zhenxing Jiajiang zaozhide jianyi," 2.

8. Interview Shi Qinglin 27/04/96.

9. In China, physical stature is a good indicator of wealth and status. Rural men in Sichuan are 8 cm shorter than their urban counterparts and 10 cm shorter than the national urban average. The comparable figures for women are 5 and 7 cm. City people tend to be proud of their "better" physique, and associate the shorter stature of migrants with poor "physical," "mental," and "moral" qualities (*shenti suzhi, sixiang daode suzhi*) (see Liu Hongkang, *Zhongguo renkou: Sichuan fence*, 364, 359–61).

10. Interviews Shi Qingcheng 06/10/95, Peng Chunbin 11/11/95.

11. Liu Shaoquan, *Jiajiang de zhiye yu guoji jiaoliu*, 36; Zhongguo renmin yinhang, "Jiajiang zhiye shengchan jingying."

12. Guojia tongjiju, *Zhongguo nongcun tongji nianjian*, 1996, 329.

13. Christine Wong, "Central-Local Relations," 694.

14. Ibid., 711.

15. Like several other successful paper traders, Li was tainted by bad class origin. His father had been an officer in the Guomindang army and was shot as a "spy," and his mother had fled from Kaixian in eastern Sichuan to Jiajiang, where she had relatives. Like the bad-class paper traders Peng Chunbin and Shi Weifang, Li worked as a tailor for much of the collective period.

16. Acetylene is used for welding and cutting steel.

17. Interviews Shi Jianrong 01/12/95, Zhang Xuelin 04/12/95, Shi Longji 21/11/95, Shi Shengyi 18/11/95, Shi Shengyi 28/07/2001; *Leshan ribao*, "Zhixiang daitouren."

18. Interviews Shi Shengcai and Shi Jianrong 27/07/2001, Shi Shengyi 28/07/2001.

19. Other observers of TVEs in Sichuan also noted that benefits were unevenly distributed; see Ruf, *Cadres and Kin*, 141–50; and Yang Minchuan, "Reshaping Peasant Culture," 168–74.

20. *Leshan ribao*, "Zhixiang daitouren."

21. In 1994, Shiyan's workforce numbered 740, of which 130 worked in agriculture, 15 in construction, 2 in transportation, 10 as shopkeepers, and 40 in service industries and handicrafts other than papermaking. The remaining 543 were classified as "industrial labor"; of this number, 50 were registered as working in the factories, and 493 in the paper workshops (interview Shi Jianrong 1/12/95).

22. Xinghuo paid piece-rate wages, which averaged around RMB 300 a month; additional benefits (health care, bonuses, New Year presents) amounted to RMB 200 a year. The acetylene factory paid an average monthly wage of RMB 270 to shop-floor workers, for far less demanding work.

23. For different terms for work and their association with the urban-industrial vs. rural-agrarian sector, see Henderson et al., *Re-drawing Boundaries*, 39–48.

24. During my fieldwork, I lived in the office building of the acetylene factory and ate in the factory canteen.

25. Interview Shi Longji 21/11/95.

26. Interview Shi Chunjin 22/04/96.

27. Papermakers told me that artists' preference for handmade paper is "a kind of superstition" and that machine-made paper can be just as good. The problem with machine-made paper is that most of the fibers (which tend to align with the moving belt) point in the same direction and therefore are less matted than in handmade paper. In consequence, the paper tears easily, especially if the calligrapher uses a heavy, wet brush. It is, however, possible to adjust paper machines so that they produce paper with great tensile strength.

28. Interviews Yang Dehua 24/11/95, Shi Shengfu 19/07/96.

29. Zhongguo renmin yinhang, "Jiajiang zhiye shengchan jingying"; Zhang Wanshu, "Guanyu zhenxing Jiajiang zaozhide jianyi."

30. Interview Huang Fuyuan 16/07/96.

31. Zhongguo renmin yinhang, "Jiajiang zhiye shengchan jingying," 6.

32. Interview Xiao Zhicheng 25/09/98.

33. County head Ma Youmei, "Jiajiang wupo wuli." In the climate of these years, Ma's injunctions against "fear" and "bookishness" (*kao benben de jiu xiguan*) and his endorsement of "risk," "daring," and "creativity" would have been understood as an encouragement to disregard environmental protection laws and other rules that stood in the way of rapid growth.

34. "Zhengfu yao shanyu faxian he huajie shehui maodun, ba shezhong naoshi miaotou xiaochu zai mengya zhuangtai"; speech by county party secretary Li Liugen, December 1999, in Sichuan sheng Jiajiang xian bianshi xiuzhi weiyuanhui, *Jiajiang nianjian*, 1999, 138–39. For the promise of subsidies and cheap loans, see Ma Youmei, "Qiangji guben."

Chapter 9

1. I calculated production costs and profits of seven *shengxuan* and nine *jingliao* producers. Average yearly workshop income in 1995 was 5,728 *yuan* for *shengxuan* and 18,440 *yuan* for *jingliao* producers. The average profit from a *dao* (100 sheets) of *shengxuan* was 6.54 *yuan*, versus 15.16 for *jingliao*.

2. In fact, crime was rare in the village. I know of only one incident, in which a pulp machine was stolen from a locked workshop. This seems to

have been an act of revenge, perhaps resulting from a quarrel, rather than a break-in by an outsider. Perhaps not by coincidence, the owner was not a Shi, although his mother was (interview Yang Yuanjin 12/10/95).

3. To ensure the anonymity of interviewees, I have changed these generation markers.

4. Apart from the first two characters, the generation names (*ke xue kai hui liang deng chao cheng an xiang bang qian si xian yi ze ren shang yuan chang*) do not form a meaningful phrase, although all have auspicious meanings.

5. Interviews Shi Dingsheng 17/11/95, Shi Dingliang 26/09/98.

6. Wang Huning, *Dangdai Zhongguo cunluo jiazu wenhua*, 81–85; Pieke, "Genealogical Mentality," 101–5; Jing Jun, *Temple of Memories*.

7. As Andrew Kipnis (*Producing Guanxi*, 173–74) has noted, the Chinese notion of tradition (*chuantong*) has a more active connotation than the English term; it implies an obligation not only to "have children to continue patrilines, but [also to] impart to them skills, knowledge, and practices that constitute genuine generational continuity." The word used here is not *chuantong* (or the verb *chuan*, "to transmit") but *yishou* ("to bequeath and instruct"), but the implication is the same.

8. Several Shis explained to me that the widow who first settled in Mianzhupu chose the surnames Wang, Feng, and Shi to mark the birth order of her sons. Wang 汪 has three "dots," Feng 冯 has two, and Shi 石 should be written with one dot on the right-hand side (although the proper pronunciation of that character would then be *dan*). Of course, most Shis interpreted this as meaning that they were the descendants of the eldest, not the youngest, son.

9. Dikötter, *Imperfect Conceptions*, chap. 4; Anagnost, *National Past-Times*, chap. 5; Greenhalgh and Winckler, *Governing China's Population*, 117, 125, 171.

10. Hsu, "Observations on Cross-Cousin Marriage," 84–86; Cooper, "Cousin Marriage," 778–79.

11. Feng Han-Yi, "The Chinese Kinship System," 167.

12. Dikötter, *Imperfect Conceptions*, 59.

13. Interview Shi Haibo 11/04/96.

14. Kipnis, "Within and Against Peasantness."

15. Yan Yunxiang, *Private Life Under Socialism*, 220–35.

16. On the paucity of cultural content in Chinese nationalism, see Fitzgerald, "The Nationless State."

17. Pieke, "Genealogical Mentality"; Jing Jun, *Temple of Memories*; Wang Huning, *Dangdai Zhongguo cunluo jiazu wenhua*, 81–85.

18. Yan Yunxiang, *Private Life Under Socialism*, 225–26.

Conclusion

1. Marx, "The Eighteenth Brumaire of Louis Bonaparte," 608; Marx and Engels, "Manifesto of the Communist Party," 477; Engels, "From Paris to Bern," 519. For a discussion of Marx's and Engel's views on peasants, see Draper, *Karl Marx's Theory of Revolution*, 2: 317, 337–39, 344–48.

2. Eugen Weber, *Peasants into Frenchmen*, 3–7.

3. Marshall, *Citizenship and Social Class*.

4. Somers, "Rights, Relationality, and Membership," 105. For a similar argument on *ancien régime* France, see Sewell, *Work and Revolution in France*.

5. Somers, "Rights, Relationality, and Membership," 105.

6. Ibid., 105–6.

7. Brook, "Auto-Organization in Chinese Society," 22. For traditional forms of self-governance in China, see also Gamble, *North China Villages*; and Feuchtwang, "Peasants, Democracy, and Anthropology."

8. Nisbet, "Citizenship: Two Traditions," 8.

9. Faure and Liu, "Introduction," 1. See also Yi-tsi Mei Feuerwerker, *Ideology, Power, Text*; and Han Xiaorong, *Chinese Discourses on the Peasant*, 19–40.

10. Bell, *One Industry, Two Chinas*; Honig, *Sisters and Strangers*; Hershatter, *The Workers of Tianjin*; Perry, *Shanghai on Strike*.

11. Perry, "Introduction." For "cultures of attachment," see Somers, "Rights, Relationality, and Membership," 105.

12. Perry, "From Native Place to Workplace."

13. Eugen Weber, *Peasants into Frenchmen*, 9.

14. Coronil, "Smelling Like a Market," 119–20.

15. Ibid., 126–27.

16. Mueggler, *The Age of Wild Ghosts*, chap. 6; idem, "Poetics of Grief," 984–91.

17. For an (in my view unconvincing) argument to the contrary, see S. A. Marglin, "What Do Bosses Do?"; and idem, "Understanding Capitalism." For a critique of Marglin, see Landes, "What Do Bosses Really Do?" See also Berg, "On the Origin of Capitalist Hierarchy"; and Thompson, *Nature of Work*.

18. Sabel, *Work and Politics*; Wood, *Degradation of Work*.

19. James Scott, *Seeing Like a State*; Mitchell, *Rule of Experts*; F. A. Marglin and S. A. Marglin, *Dominating Knowledge*; Stone, *Agricultural Deskilling*.

20. Burawoy and Krotov, "The Soviet Transition from Socialism to Capitalism."

21. Mokyr, *Gifts of Athena*; Sabel and Zeitlin, *Worlds of Possibilities*; idem, "Historical Alternatives to Mass Production"; Piore and Sabel, *Second Industrial Divide*; Berg, "Factories, Workshops, and Industrial Organisation." See also Morris-Suzuki, *Technological Transformation of Japan*.

22. This is partly a semantic issue: *jishu*, the most common word for skill, is associated with skilled work in the urban sector, as in the term "skilled worker," *jishu gongren*.

23. Yan Hairong, "Neoliberal Govermentality."

24. Pun, *Made in China*, 115–16.

Works Cited

Sichuan Journals

SCJJJK 1944 [1: 3]. "Yinianlai Chuansheng mijia biandong zhi huigu" (Looking back on one year of rice price fluctuations in Sichuan). *Sichuan jingji jikan* 1, no. 3 (Jun. 1944): 276–86.

SCJJJK 1945 [2: 2]. "Sichuan jingji tongji" (Economic statistics of Sichuan). *Sichuan jingji jikan* 2, no. 2 (Apr. 1945): 332–61.

SCJJJK 1946 [3: 3]. "Sichuan sheng ge xian shi huliang xingbili he renkou midu" (Household size, sex ratios, and population density in the counties and cities of Sichuan). *Sichuan jingji jikan* 3, no. 3 (Aug. 1946): 119–22.

SCJJYK 1935 [3: 1]. "Jiajiang gailiang zhiye" (Jiajiang reforms its paper industry). *Sichuan jingji yuekan* 3, no. 1 (Jan. 1935): 187–88.

SCJJYK 1935 [3: 2]. "Jiajiang zhiye diaocha" (Survey of the Jiajiang paper industry). *Sichuan jingji yuekan* 3, no. 2 (Feb. 1935): 89–90.

SCJJYK 1935 [3: 3]. "Liangshan zhiye gaikuang" (General conditions in the paper industry of Liangshan). *Sichuan jingji yuekan* 3, no. 3 (Mar. 1935): 119–21.

SCJJYK 1935 [3: 4–5]. "Jianting shishi zhizhi gongye sinian jihua" (The Reconstruction Bureau implements a four-year plan for the paper industry). *Sichuan jingji yuekan* 3, no. 4–5 (Apr.–May 1935): 145–48.

SCJJYK 1935 [3: 4–5]. "Liang-Da jianshe jiqi zaozhichang" (Liangshan and Daxian counties construct mechanized paper factory). *Sichuan jingji yuekan* 3, no. 4–5 (Apr.–May 1935): 148–50.

SCJJYK 1935 [4: 2]. "Liangshan zaozhiye gailiang chanpin" (The paper industry of Liangshan reforms its products). *Sichuan jingji yuekan* 4, no. 2 (July 1935): 119.

SCJJYK 1936 [5: 2–3]. "Jiajiang gaikuang" (General conditions in Jiajiang). *Sichuan jingji yuekan* 5, no. 2–3 (Feb.–Mar. 1936): 112–14.

SCJJYK 1936 [5: 4]. "Jiading zhiye xiankuang" (The current situation of the paper industry of Jiading [Leshan]). *Sichuan jingji yuekan* 5, no. 4 (April 1936): 19–21.

SCJJYK 1936 [6: 6]. "Zhiye jinxun" (Latest news on the paper industry). *Sichuan jingji yuekan* 6, no. 6 (Dec. 1936): 41–43.

SCYB 1932 [1: 1]. "Zhizaoye" (Manufacture). *Sichuan yuebao* 1, no. 1 (July 1932): 19–21.

SCYB 1932 [1: 2]. "Liangshan huangbiao zhiye" (*Huangbiao* paper production in Liangshan county). *Sichuan yuebao* 1, no. 2 (Aug. 1932): 29.

SCYB 1932 [1: 3]. "Gedi zaozhiye gaikuang" (Description of the paper industry in different localities). *Sichuan yuebao* 1, no. 3 (Sept. 1932): 58–60.

SCYB 1932 [1: 4]. "21 jun fuzhu Liang-Da zhiye" (The 21st Army supports the paper industry in Liangshan and Dazhu counties). *Sichuan yuebao* 1, no. 4 (Oct. 1932): 60–61.

SCYB 1933 [2: 2]. "Guang'an xian zhi zaozhi gongye" (The paper industry of Guang'an county). *Sichuan yuebao* 2, no. 2 (Feb. 1933): 49–50.

SCYB 1933 [2: 4]. "Quan Chuan zhichan gaikuang" (Brief description of paper production in all Sichuan). *Sichuan yuebao* 2, no. 4 (Apr. 1933): 149–50.

SCYB 1933 [3: 2]. "Sichuan zhi zhiye" (The paper industry of Sichuan). *Sichuan yuebao* 3, no. 3 (July 1933): 1–20.

SCYB 1934 [4: 2]. "Leshan Jiale zhichang kuochong jihua" (Plans for the enlargement of the *Jiale* paper factory). *Sichuan yuebao* 4, no. 2 (Feb. 1934): 97–98.

SCYB 1934 [4: 2]. "Tongliang gailiang zhichang ji cichang" (Improved paper and porcelain factories in Tongliang county). *Sichuan yuebao* 4, no. 2 (Feb. 1934): 98.

SCYB 1934 [4: 3]. "Sichuan zaozhi yuanliao diaocha" (Survey of the raw materials for Sichuan's paper industry). *Sichuan yuebao* 4, no. 3 (Mar. 1934): 112–17.

SCYB 1934 [4: 5]. "Guang'an zhi zaozhi gongye" (The paper industry of Guang'an). *Sichuan yuebao* 4, no. 5 (May 1934): 100–104.

SCYB 1934 [4: 5]. "Leshan Jiale zhichang gaikuang" (Description of the Jiale paper factory in Leshan). *Sichuan yuebao* 4, no. 5 (May 1934): 104–9.

SCYB 1934 [5: 1]. "Sichuan chanzhi quyu gaikuang" (Situation in the paper production districts of Sichuan). *Sichuan yuebao* 5, no. 1 (July 1934): 130–31.

SCYB 1934 [5: 6]. "Jiajiang zhizhi gongye gaikuang" (Description of the paper industry in Jiajiang). *Sichuan yuebao* 5, no. 6 (Dec. 1934): 155–61.

SCYB 1935 [6: 1]. "Jiajiang qing mianzheng 24 nian liangshui" (Jiajiang asks for waiver of 1935 grain tax). *Sichuan yuebao* 6, no. 1 (Jan. 1935): 17.

SCYB 1935 [6: 2]. "Jiaijang zhiye diaocha" (Survey of Jiajiang papermaking). *Sichuan yuebao* 6, no. 2 (Feb. 1935): 77–79.

SCYB 1935 [6: 2]. "Liangshan Dazhu zuzhi lianhe zhizhichang" (Liangshan and Dazhu counties organise a joint paper factory). *Sichuan yuebao* 6, no. 2 (Feb. 1935): 141.

SCYB 1935 [6: 3]. "Shengfu shixing zhenxing Chuansheng zaozhi gongye jihua" (Plan of the provincial government for the revival of Sichuan's paper industry). *Sichuan yuebao* 6, no. 3 (Mar. 1935): 159–60.

SCYB 1935 [7: 3]. "Liangshan chouban gailiang zhichang" (Liangshan plans an improved paper factory). *Sichuan yuebao* 7, no. 3 (Sept. 1935): 151–52.

SCYB 1935 [7: 3]. "Shengfu ming gailiang Jiale zhi" (The provincial government orders the reform of Jiale paper). *Sichuan yuebao* 7, no. 3 (Sept. 1935): 150–51.

SCYB 1935 [7: 5]. "Dazhu chouban Minsheng zaozhichang" (Dazhu county plans a Minsheng paper factory). *Sichuan yuebao* 7, no. 5 (Nov. 1935): 136–38.

SCYB 1936 [8: 6]. "Jiajiang zhiye pochan" (The bankruptcy of Jiajiang papermaking). *Sichuan yuebao* 8, no. 6 (June 1936): 101–2.

SCYB 1936 [9: 1]. "Jianting diaocha Jia-Hong-E sanxian zhiye jinkuang" (Reconstruction Bureau examines the recent situation of the paper industry in Jiajiang, Hongya, and Emei). *Sichuan yuebao* 9, no. 1 (July 1936): 121–23.

SCYB 1936 [9: 1]. "Sichuan sheng zhi gongye" (The industries of Sichuan). *Sichuan yuebao* 9, no. 1 (July 1936): 242–51.

SCYB 1937 [10: 2]. "Guang'an choushe da zhichang" (Guang'an county plans a big paper factory). *Sichuan yuebao* 10, no. 2 (Feb. 1937): 184.

SCYB 1937 [10: 3]. "Tongliang techan diaocha" (Description of the special products of Tongliang). *Sichuan yuebao* 10, no. 3 (Mar. 1937): 178–88.

SCYB 1937 [10: 5]. "4 qu zhuanshu nizai Jiajiang choushe zhichang" (Administrative Office of 4th District plans to establish paper factory in Jiajiang). *Sichuan yuebao* 10, no. 5 (May 1937): 217–18.

SP 1925. "Sichuan Jiajiang xian zhi zhiye" (The paper industry of Jiajiang in Sichuan). *Shuping* 4, no. 3 (Mar. 1925): 35–40.

Sichuan sheng zhengfu gongbao 1936. "Wei lingzhi gaixian chafu meinian zaozhi zhonglei, shuliang, chengben fei, ji shoujia yi an ling yang zunzhao you" (Concerning the order to the counties to report yearly on paper sorts, amounts, capital costs, and sales prices). *Sichuan sheng zhengfu gongbao* 50 (July 1936): 21–22.

Documents from the Sichuan Archives

Gongyeting 1951 [13]. "Chuanxiqu zaozhi gongye de jiben qingkuang ji cunzai wenti" (The basic condition of papermaking in West Sichuan and its problems).

Gongyeting 1951 [19: 1]. West Sichuan Administration, Industry Bureau. "Xi'nan di'erjie zaozhi huiyi zongjie" (Summary of the second Southwest paper industry meeting).

Gongyeting 1951 [19: 2]. West Sichuan Administration, Industry Bureau. "Chuanxiqu zaozhi gongye yuanliao diaocha zongjie baogao" (Summary report on the survey on raw materials for the paper industry of West Sichuan district).

Gongyeting 1951 [19: 3]. West Sichuan Administration, Industry Bureau. "Chuanxiqu shoujie zhiye huiyi zongjie" (Summary of the first meeting of the paper industry of the West Sichuan district).

Gongyeting 1951 [19: 4a]. West Sichuan Administration, Industry Bureau. "Chuanxiqu shoujie zhiye huiyi ziliao" (Materials on the first meeting of the paper industry of the West Sichuan district).

Gongyeting 1951 [19: 4b]. West Sichuan Administration, Industry Bureau. "Xi'nan di'erjie zaozhi huiyi zongjie" (Summary of the second Southwest paper industry meeting).

Gongyeting 1951 [19: 5]. West Sichuan Administration, Industry Bureau. "Chuanxiqu diyijie zaozhi huiyi choubei tigang" (Preparatory outline of the first meeting of the paper industry of the West Sichuan district).

Gongyeting 1951 [93: 1]. West Sichuan Administration, Industry Bureau. "Di'erjie Xi'nanqu zaozhi gongye huiyi guanyu fudao zuzhi shougong zhiye de jueyi" (Decisions regarding guidance and organization of handicraft paper production adopted at the 2nd Southwest paper industry meeting).

Gongyeting 1951 [93: 2]. West Sichuan Administration, Industry Bureau. "Xi'nanqu shougong zaozhiye fudao weiyuanhui zuzhi guicheng— xiuzheng cao'an" (Organizational statute for the guiding committee for the southwest handicraft paper industry—revised draft version).

Gongyeting 1951 [93: 3]. West Sichuan Administration, Industry Bureau. "Xi'nanqu shougong zaozhiye fudao weiyuanhui zuzhi fudao shishi banfa" (Methods of guidance and organization for the guidance committee of the southwest handicraft paper industry).

Gongyeting 1951 [93: 6]. West Sichuan Administration, Industry Bureau. "Di'erjie Xi'nanqu shougong zaozhiye fudao weiyuanhui" (Second meeting of the guidance committee of the southwest handicraft paper industry).

Gongyeting 1951 [146: 1]. West Sichuan Administration, Industry Bureau. "Chuanxiqu gequ zhi chan-xiao tongji" (Local data on paper output and sales in the West Sichuan district).

Gongyeting 1951 [171: 1]. "Jiajiang zhiye diaocha baogao" (Survey report on the paper industry of Jiaijang).

Gongyeting 1951 [171: 2]. "Chuanxi gongyeting shougongye ziliao: Mianzhu xian shougong zaozhiye diaocha baogao (baokuo Shenfang Sanhe xiang, Maoxian Qingping xiang)" (West Sichuan Industry Bureau handicraft materials: survey report on the handicraft paper industry of Mianzhu county, including Sanhe township in Shenfang and Qingping township in Maoxian county).

Gongyeting 1951 [171: 3]. "Chuanxi gongyeting shougongye ziliao: Chongqing xian de zaozhiye" (West Sichuan Industry Bureau handicraft materials: the handicraft paper industry of Chongqing county).

Gongyeting 1951 [171: 4]. "Chuanxi gongyeting shougongye ziliao: Dayi xian de zaozhiye" (West Sichuan Industry Bureau handicraft materials: the handicraft paper industry of Dayi county).

Gongyeting 1951 [171: 5]. "Chuanxi gongyeting shougongye ziliao: Chuanxi qu zaozhi gongye yuanliao diaocha zongjie baogao" (West Sichuan Industry Bureau handicraft materials: Summary report on the survey on raw materials for the paper industry of West Sichuan).

Gongyeting 1952 [146: 2]. West Sichuan Administration, Industry Bureau. "Jiajiang shifan zhichang gaikuang" (Description of the experimental paper factory in Jiajiang).

Gongyeting 1952 [146: 3]. West Sichuan Administration, Industry Bureau. "Chuanxi Jiajiang shifan zhichang kuiben de yuanyin ji changzhong qingkuang" (Reasons for the losses incurred by the experimental paper factory of Jiajiang and description of the factory).

Gongyeting 1952 [106]. West Sichuan Administration, Industry Bureau. "Chuanxiqu qing erye zhongdian hangye 1951 nian nianzhong zongjie" (Year-end summary of the key branches of the second light industry in 1951).

Jianchuan 1960 [80-3111]. Light Industry Ministry. "Sichuan sheng qinggongye ting baosong zaozhi gongzuo shangbannian zongjie he xiabannian gongzuo anpai yijian" (The Light Industry Bureau of Sichuan sends a summary report on paper production work in the first half of the year and work plans and suggestions for the second half of the year).

Jianchuan 1963 [074-17]. Provincial Government of Sichuan. "Zhonggong Sichuan shengwei pizhuan Miao Fengshu tongzhi zai sheng shougongye gongzuohuiyi shang de baogao" (Party Committee of Sichuan province approves report by Comrade Miao Fengshu at the handicraft work meeting).

Jianchuan 1963 [074-99b]. Provincial Committee for Industrial Production. "Guanyu zhaokai sheng shougongye gongzuo huiyi de qingkuang baogao" (Report on conditions for convening a work meeting on handicrafts in the province).

Jianchuan 1963 [074-99e]. Provincial Handicraft Bureau. "Sichuan sheng shougongye tiaozheng gongzuo jinzhan qingkuang jianbao" (Brief report on the progress of the adjustment work in Sichuan's handicrafts).

Jiansheting 1936 [1353a: 1]. Reconstruction Bureau. "Ling diaocha shengtan xiaoshou Jiajiang, Hongya, Emei sanxian zhizhang gaikuang" (Order to register sales within the province of paper from the three counties of Jiajiang, Hongya, and Emei).

Jiansheting 1936 [1353a: 2]. Jiajiang county government. "Jiajiang xian chuchan zhizhang zhonglei shuliang ji chengben xiaojia zongshu yilanbiao" (Table of categories, amounts, costs, and prices of paper produced in Jiajiang county).

Jiansheting 1936 [1353a: 3]. Letter by Jiajiang county magistrate Du Ao accompanying [1353a: 2].

Jiansheting 1936 [1353a: 4]. Emei county government. "Emei xian meinian chuchan zhizhang gaikuang" (Yearly paper production in Emei).

Jiansheting 1936 [1353a: 5]. Hongya county government. "Hongya xian zhizhang zhonglei shuliang ji chengben xiaojia zongshu yilanbiao" (Table of categories, amounts, costs, and prices of paper produced in Hongya county).

Jiansheting 1937 [1353a: 6]. Huang Yonghai. "Chengwei tuzhi pochan kenqing xun yu jiuji yi wei nongmin fuchan er fuxing nongcun you" (Petition concerning the bankruptcy of paper production, asking for aid to support farm sideline production and revive the countryside).

Jiansheting 1937 [1353a: 7]. Headquarters of the Chairman of the National Defense Committee. "Ling Sichuan shengfu: ju Jiajiang zhiye gonghui Huang Yonghai deng cheng wei tuzhi pozhan, kenqing xunyu jiuji yi wei nongmin fuchan er fuxing nongcun dengqing lingyang xun ni jiuji banfa chenghe you" (Order to the Sichuan provincial government: concerning the petition by the chairman of the Jiajiang paper guild Huang Yonghai "concerning the bankruptcy of paper production, asking for aid to support farm sideline production and revive the countryside": we order you to speedily devise a way of solving the problem and report).

Jiansheting 1937 [1353a: 8, 9, 10, 11]. Reconstruction Bureau. "Sichuan sheng jiantiao" (Sichuan province, Reconstruction Bureau, drafts of orders and letters to the Finance Bureau, the provincial government, and Jiajiang county).

Jiansheting 1937 [1353a: 12]. Sichuan provincial government. "Sichuan sheng zhengfu xunling ling benfu . . ." (The provincial government of Sichuan orders its Reconstruction Bureau . . .).

Jiansheting 1937 [1353a: 13]. Reconstruction Bureau. "Wei ju benshi zhizhang yinshuaye tongye gonghui chengbao zhizhang quefa qingxing" (Concerning the petition of the Chongqing paper and printing guild on the paper shortage).

Jiansheting 1938 [9337: 1]. Jiajiang county govenrment. "Shiyou: chengbao zhixing xian xingzheng huiyi juean qing pai yuan gailiang zhiye" (Re: Report on the implementation of a decision of the county's administrative conference, petition to send staff to reform the paper industry).

Jiansheting 1938 [9337: 2]. Mianzhu county government. "Shiyou: wei cixian zhiye jiying gailiang, niqing junfu lingpai jishu zhuanjia" (Re: The paper industry of this county being in urgent need of reform, we ask you to send technical experts).

Jiansheting 1939 [4042: 1]. *Chengdu kuaibao, Huaxi ribao, Xinxin xinwen*, et al. "Chengwei zhishang jianshang baocong juji qingyu tongzhi yiwei shehui wenhua shi" (Petition to control the unscrupulous profiteering of paper and soda merchants to safeguard society and culture).

Jiansheting 1939 [4042: 2]. Chengdu city government. "Shiyou: chengfu zunban jiejiu benshi xinwen zhihuang qingxing qing yu jianhe shi zun you" (Re: Responding to the order to solve the paper scarcity of this city and examine the situation).

Jiansheting 1939 [9337: 5a]. Jiajiang county government. "Zunling chengbao gaijin zhiye banfa, ji quti mixin yongzhi qingxing" (Having received your order, we report on the reform of the paper industry and the substitution of superstition paper).

Jiansheting 1939 [9337:5b]. Reconstruction Bureau. "Zunjing chengbao gaijin zhi" (Order to undertake reform of paper production).

Jiansheting 1941 [1353b: 2]. Reconstruction Bureau. "Ling Tongliang, Du'an, Dazhu, Liangshan . . ." (Order to the county governements of Tongliang, Du'an, Dazhu, Liangshan . . .).

Jiansheting 1942 [1353b: 1]. Reconstruction Bureau. "Sichuan sheng zheng-fu ling Yongchuan, Tongliang, . . ." (Order by the Sichuan Bureau of Reconstruction to the counties of Yongchuan, Tongliang, . . .).

Jiansheting 1942 [1353b: 3]. "Jiajiang zhi shuomingbiao" (Explanatory table of Jiajiang paper).

Jiansheting 1942 [9338: 1]. Wang Yunming, Jiajiang county magistrate. "Shiyou: wei quan xian wei zaozhi gongyequ shan duo tian shao . . ." (Re: The entire county is a paper-producing industrial district, mountains are many and fields are few, . . .).

Jiansheting 1942 [9338: 2]. Finance Bureau, Land Tax Office. "Shiyou: zhun han zhuhe ban Jiajiang xianzhang Wang Yunming qing jianwei miangou" (Re: Your letter, concerning the request of the Jiajiang magis-trate Wang Yunming for a reduction of taxes and an exemption from purchases).

Jiansheting 1942 [9338: 3]. Jiajiang county. "Wei niju zhiye gaijin jihua qing zhuan Nongmin yinhang daikuan" (Re: In order to carry out the planned reform of paper production, we request a loan from the Agricul-tural Bank).

Jiansheting 1942 [9338: 4]. Jiajiang county. "Jiajiang xian zhiye gaijin ji-huashu" (Reform plans for the Jiajiang paper industry).

Jiansheting 1943 [9338: 5]. National Ministry of Economy. "Anju benbu Riyong bixupin guanlichu . . ." (Concerning the Office for the Man-agement of Daily Necessities . . .).

Jiansheting 1945 [5117: 1]. Jiajiang Paper Production and Marketing Co-operative. "Baozheng zeren Jiajiang zhiye shengchan yunxiao hezuoshe gaikuang baogao" (Summary report on the carrying out of responsibili-ties of the Jiajiang Paper Production and Marketing Cooperative).

Jiansheting 1945 [5117: 2]. Jiajiang Paper Production and Marketing Co-operative. "Baozheng zeren Jiajiang zhiye shengchan yunxiao hezuoshe diyi niandu yewu jihua" (Plan for carrying out the responsibilities of the Jiajiang Paper Production and Marketing Cooperative in its first year).

Jiansheting 1945 [9338: 6]. National Ministry of the Economy. "Ju benbu riyong bixupin guanlichu cheng guanyu quti mixin yongzhi . . ." (Office for the Management of Daily Necessities: petition concerning the re-placement of superstition paper . . .).

Jiansheting 1948 [1307]. Jiajiang county government. "Shiyou: wei zunling tianbao difang jingji gaikuang diaochabiao ji jingji shiye fenye gaikuang diaochabiao yangqi" (Re: Carrying out the order to supplement eco-nomic survey data, we submit survey data on industrial branches).

JJ-Mucheng 1947. Jiajiang county government. "Jiajiang xian Mucheng xiang renkou tiaojie youguan ziliao" (Data on population adjustments in Mucheng township, Jiajiang county).

JJ-Shenghang 1944. Bank of Sichuan, Jiajiang Branch Office. "Yingye gai-kuang biao" (Enterprise data forms).

JJ-tax 1947. Jiajiang County Tax Office. "Bugao" (Proclamation).

JJ-tax 1949. Jiajiang County Tax Office. "Jiajiang xian shuijuan caizheng-chu: 38 nian xiabannian meiyue yingyueshui tanke" (Jiajiang County Tax Office: monthly assessments of income tax for the second half of 1949).

Sichuan hezuo jinku 1941. "Jiajiang xian hezuo jinku 30 niandu yewu bao-gaoshu" (1941 work report of the cooperative funds of Jiajiang county).

Sichuan hezuo jinku 1943. "Jiajiang xian hezuo jinku 32 niandu yewu bao-gaoshu" (1943 work report of the cooperative funds of Jiajiang county).

Secondary Sources

Alitto, Guy. The Last Confucian: Liang Shu-ming and the Chinese Dilemma of Modernity. Berkeley: University of California Press, 1979.

American Rural Small-Scale Industry Delegation. Rural Small-Scale Industry in the People's Republic of China. Berkeley: University of California Press, 1977.

Anagnost, Ann. "The Corporeal Politics of Quality (Suzhi)." Public Culture 16, no. 2 (Spring 2004): 189–208.

———. National Past-Times: Narrative, Representation, and Power in Modern China. Durham, NC: Duke University Press, 1997.

———. "Socialist Ethics and the Legal System." In Popular Protest and Po-litical Culture in Modern China: Learning from 1989, ed. Jeffrey Wasser-strom and Elizabeth J. Perry. Boulder: Westview, 1992, 177–205.

Appadurai, Arjun. "Introduction: Commodities and the Politics of Value." In The Social Life of Things: Commodities in Cultural Perspective, ed. idem. Cambridge: Cambridge University Press, 1986, 6–63.

Aristotle. Politics: Books I and II. New York: Clarendon, 1995.

Averill, Stephen C. "The Shed People and the Opening of the Yangzi High-lands." Modern China 9, no. 1 (Jan. 1983): 84–126.

Bailes, Kendall E. "Alexei Gastev and the Soviet Controversy over Taylor-ism, 1918–1924." Soviet Studies 29, no. 3 (July 1977): 373–94.

Bailey, Paul. Strengthen the Country and Enrich the People: The Reform Writ-ings of Ma Jianzhong. Richmond, Eng.: Curzon, 1998.

Baker, Hugh D. R. Chinese Family and Kinship. New York: Columbia Uni-versity Press, 1979.

Banister, Judith. *China's Changing Population*. Stanford: Stanford University Press, 1987.

Bell, Lynda S. *One Industry, Two Chinas: Silk Filatures and Peasant-Family Production in Wuxi County, 1865–1937*. Stanford: Stanford University Press, 1999.

Berg, Maxine. "Factories, Workshops, and Industrial Organisation." In *The Economic History of Britain Since 1700*, ed. Roderick Floud and Deirdre McCloskey. Cambridge: Cambridge University Press, 1994, 123–50.

———. "On the Origin of Capitalist Hierarchy." In *Power and Economic Institutions*, ed. Bo Gustafsson. Aldershot: Elgar, 1991, 173–94.

Bian, Morris L. *The Making of the State Enterprise System in Modern China: The Dynamics of Institutional Reform*. Cambridge: Harvard University Press, 2005.

Billeter, Jean-François. "The System of 'Class Status.'" In *The Scope of State Power in China*, ed. Stuart R. Schram. London: School of Oriental and African Studies, 1985, 127–69.

Bourdieu, Pierre. *Algérie 60—structures économiques et structures temporelles*. Paris: Minuit, 1977.

———. *The Logic of Practice*. Stanford: Stanford University Press, 1980.

———. *Outline of a Theory of Practice*. Cambridge: Cambridge University Press, 1977.

Bramall, Chris. *In Praise of Maoist Economic Planning: Living Standards and Economic Development in Sichuan Since 1931*. Oxford: Clarendon Press, 1993.

Braverman, Harry. *Labor and Monopoly Capital: The Degradation of Work in the Twentieth Century*. New York: Monthly Review Press, 1974.

Bray, Francesca. *Technology and Gender: Fabrics of Power in Late Imperial China*. Berkeley: University of California Press, 1997.

Brewer, John, and Roy Porter, eds. *Consumption and the World of Goods*. London: Routledge, 1993.

Brokaw, Cynthia J. *Commerce in Culture: The Sibao Book Trade in the Qing and Republican Periods*. Cambridge: Harvard University Asia Center, 2006.

Brook, Timothy. "Auto-Organization in Chinese Society." In *Civil Society in China*, ed. idem and B. Michael Frolic. Armonk, NY: M. E. Sharpe, 1997, 19–45.

Brown, Jeremy. "From Resisting Communists to Resisting America: Civil War and Korean War in Southwest China, 1950–51." In *Dilemmas of Victory: The Early Years of the People's Republic of China*, ed. idem and Paul G. Pickowicz. Cambridge: Harvard University Press, 2008, 105–29.

Burawoy, Michael. *Manufacturing Consent: Changes in the Labor Process Under Monopoly Capitalism*. Chicago: University of Chicago Press, 1979.

Burawoy, Michael, and Pavel Krotov. "The Soviet Transition from Socialism to Capitalism: Worker Control and Economic Bargaining in the Wood Economy." *American Sociological Review* 57 (Feb. 1992): 16–38.

Burawoy, Michael, and János Lukász. *The Radiant Past: Ideology and Reality in Hungary's Road to Capitalism*. Chicago: Chicago University Press, 1992.

Burgess, John Stuart. *The Guilds of Peking*. Taipei: Chengwen, 1928.

Cao Shuji. "1958–61 nian Sichuan renkou siwang" (Deaths among Sichuan's population in 1958–61). *Zhongguo renkou kexue*, Jan. 2004, 57–67.

Cao Tiansheng. *Zhongguo Xuanzhi* (Chinese Xuan paper). Beijing: Zhongguo qinggongye chubanshe, 1992.

Cartier, Carolyn. "Origins and Evolution of a Geographical Idea: The Macroregion in China." *Modern China*, 28, no. 1 (Jan. 2002): 79–142.

Chan, Anita, and Jonathan Unger. "Grey and Black: The Hidden Economy of Rural China." *Pacific Affairs* 55, no. 3 (Fall 1982): 452–71.

Chan, Anita; Richard Madsen; and Jonathan Unger. *Chen Village: The Recent History of a Peasant Community in Mao's China*. Berkeley: University of California Press, 1984.

———. *Chen Village Under Mao and Deng*. Berkeley: University of California Press, 1992.

Chan Kam-Wing. *Cities with Invisible Walls*. New York and Oxford: Oxford University Press, 1994.

Chan Kam Wing and Will Buckingham. "Is China Abolishing the *Hukou* System?" *China Quarterly*, no. 195 (2008): 582–606.

Chao Kang. *The Development of Cotton Textile Production in China*. Cambridge: East Asian Research Center, Harvard University, 1977.

———. "The Growth of a Modern Cotton Textile Industry and the Competition with Handicrafts." In *China's Modern Economy in Historical Perspective*, ed. Dwight H. Perkins. Stanford: Stanford University Press, 1975, 167–202.

Chao Yuan Ren. "Chinese Terms of Address." *Language* 32, no. 1 (1956): 217–241.

Chayanov, Aleksandr N. *The Theory of Peasant Economy*. Homewood, IL: R. D. Irwin, 1966.

Ch'en, Jerome C. *The Highlanders of Central China: A History, 1895–1937*. Armonk, NY: M. E. Sharpe, 1992.

Chen Zhongping. "The Origins of Chinese Chambers of Commerce in the Lower Yangzi Region." *Modern China* 27, no. 2 (April 2001): 155–201.

Cheng Quan (pseudonym). "Jiajiang zhishi" (A history of Jiajiang paper). Jiajiang, photocopied manuscript, n.d.

Cheng, Tiejun, and Mark Selden. "The Origins and Social Consequences of China's *Hukou* System." *China Quarterly*, no. 139 (1994): 644–68.

Chun, Allen. "The Lineage-Village Complex in Southeastern China: A Long Footnote in the Anthropology of Kinship." *Current Anthropology* 37, no. 3 (June 1996): 429–50.

Clark, Andy. *Being There: Putting Brain, Body, and World Together Again.* Cambridge: MIT Press, 1999.

Cockburn, Cynthia. *Brothers: Male Dominance and Technological Change.* London: Pluto, 1991.

Cohen, Myron L. "Cultural and Political Inventions in Modern China: The Case of the Chinese 'Peasant.'" *Daedalus* 122, no. 2 (Spring 1993): 151–70.

———. *House United, House Divided: The Chinese Family in Taiwan.* New York: Columbia University Press, 1976.

———. *Kinship, Contract, Community, and State: Anthropological Perspectives on China.* Stanford: Stanford University Press, 2005.

———. "Lineage Organization in North China." *Journal of Asian Studies* 49, no. 3 (1990): 509–34.

Cooper, Eugene. "Cousin Marriage in Rural China: More and Less than Generalized Exchange." *American Ethnologist* 20, no. 4 (Nov. 1993): 758–80.

Cooper, Eugene, with Jiang Yinhuo. *The Artisans and Entrepreneurs of Dongyang County: Economic Reform and Flexible Production in China.* Armonk, NY: M. E. Sharpe, 1998.

Coronil, Fernando. "Smelling Like a Market." *American Historical Review* 106, no. 1 (Feb. 2001): 119–29.

Davis, Richard L. "Political Success and the Growth of Descent Groups: The Shih of Ming-Chou During the Song." In *Kinship Organization in Late Imperial China, 1000–1940,* ed. Patricia Ebrey and James L. Watson. Berkeley: University of California Press, 1986, 62–94.

Dikötter, Frank. *Imperfect Conceptions: Medical Knowledge, Birth Defects and Eugenics in China.* London: Hurst, 1998.

Domenach, Jean-Luc. *The Origins of the Great Leap Forward: The Case of One Chinese Province.* Boulder, CO: Westview, 1995.

Donham, Donald L. *Marxist Modern: An Ethnographic History of the Ethiopian Revolution.* Berkeley: University of California Press, 1999.

Donnithorne, Audrey. "Sichuan's Agriculture: Depression and Revival." *Australian Journal of Chinese Affairs,* no. 12 (July 1984): 59–86.

Douglas, Mary, and Baron Isherwood. *The World of Goods: Towards an Anthropology of Consumption*. London: Lane, 1979.

Draper, Hal. *Karl Marx's Theory of Revolution*. 4 vols. New York: Monthly Review Press, 1977–90.

Dreyfus, Hubert L. *Being-in-the-World: A Commentary of Heidegger's Being and Time, Division 1*. Cambridge: MIT Press, 1991.

Du Shihua. "Shougong zhujiang de zhizao jiqi gaijin fangfa" (Manual production of bamboo pulp and how to reform it). *Zaozhi gongye*, no. 7 (1957): 25–29.

Duan Zhiyi. "Sichuan shougong zaozhiye de jishu gailiang" (Technical reform in Sichuan's handicraft paper industry). *Zhongguo gongye*, no. 12 (Oct. 1943): 37–40.

Duara, Prasenjit. *Culture, Power, and the State: Rural North China, 1900–1942*. Stanford: Stanford University Press, 1988.

Durkheim, Emile. *The Division of Labor in Society*. New York: Free Press, 1984.

Eastman, Lloyd. *Fields, Families, and Ancestors: Constancy and Change in China's Social and Economic History, 1550–1945*. Oxford: Oxford University Press, 1988.

———. *Seeds of Destruction*. Stanford: Stanford University Press, 1984.

Ebrey, Patricia. "Early Stages of Descent Group Organization." In *Kinship Organization in Late Imperial China, 1000–1940*, ed. idem and James L. Watson. Berkeley: University of California Press, 1986, 16–61.

Ebrey, Patricia, and James L. Watson. "Introduction." In *Kinship Organization in Late Imperial China, 1000–1940*, ed. idem. Berkeley: University of California Press, 1986, 1–15.

Edwards, Richard. *Contested Terrain: The Transformation of the Workplace in the Twentieth Century*. London: Heinemann, 1979.

Emerson, John Philip. *Non-Agricultural Employment in Mainland China, 1949–1958*. Washington, DC: U.S. Bureau of the Census, 1965.

Endicott, Stephen. *Red Earth*. London: Tauris, 1988.

Engels, Frederick. "From Paris to Berne." In *Karl Marx and Frederick Engels: Collected Works*. New York: International Publishers, 1975, 7: 511–29.

Entenmann, Robert. "Sichuan and Qing Migration Policy." *Qingshi wenti* 4, no. 4 (1980): 35–54.

Entwistle, Barbara, and Gail Henderson, eds. *Re-drawing Boundaries: Work, Household, and Gender in China*. Berkeley: University of California Press, 2000.

Eyferth, Jacob. "De-industrialization in the Chinese Countryside: Handicrafts and Development in Jiajiang (Sichuan), 1935–1978." *China Quarterly*, no. 173 (March 2003): 53–72.

———. "Introduction." In *How China Works: Perspectives on the Twentieth-Century Workplace*, ed. idem. Milton Park, Eng.: Routledge, 2006, 1–24.

Faure, David. *The Structure of Chinese Rural Society: Lineage and Village in the Eastern New Territories, Hong Kong*. Hong Kong: Oxford University Press, 1986.

Faure, David, and Tao Tao Liu. "Introduction." In *Town and Country in China: Identity and Perception*, ed. idem. New York: Palgrave, 2002, 1–16.

Fei Hsiao-t'ung [Fei Xiaotong]. *Peasant Life in China: A Field Study of Country Life in the Yangtze Valley*. New York: E. P. Dutton, 1936.

Fei Hsiao-t'ung [Fei Xiaotong] and Chang Chih-i [Zhang Zhiyi]. *Earthbound China: A Study of Rural Economy in Yunnan*. Chicago: University of Chicage Press, 1945.

Feng Han-Yi. "The Chinese Kinship System." *Harvard Journal of Asiatic Studies* 2, no. 2 (July 1937): 141–275.

Feuchtwang, Stephan. "Peasants, Democracy and Anthropology: Questions of Local Loyalty." *Critique of Anthropology* 23, no. 1 (2003): 93–120.

Feuerwerker, Albert. "Economic Trends, 1912–1949." In *The Cambridge History of China*, vol. 12, part 1, ed. John K. Fairbank. Cambridge, Eng.: Cambridge University Press, 1983, 28–127.

Feuerwerker, Yi-tsi Mei. *Ideology, Power, Text: Self-Representation and the Peasant "Other" in Modern Chinese Literature*. Stanford: Stanford University Press, 1998.

Fewsmith, Joseph. "From Guild to Interest Group: The Transformation of Public and Private in Late Qing China." *Comparative Studies in Society and History* 25, no. 4 (Oct. 1983): 617–40.

Fitzgerald, John. "The Nationless State: The Search for a Nation in Modern Chinese Nationalism." *Australian Journal of Chinese Affairs*, no. 33 (Jan. 1995): 75–104.

———. "Warlords, Bullies, and State Building in Nationalist China: The Guangdong Cooperative Movement, 1932–1936." *Modern China* 23, no. 4 (Oct. 1997): 420–58.

Flower, John. "Peasant Consciousness." In *Post-Socialist Peasants? Rural and Urban Constructions of Identity in Eastern Europe, East Asia, and the Former Soviet Union*, ed. Pamela Leonard and Deema Kaneff. Houndsmill, Basingstoke: Palgrave, 2002, 44–72.

Frazier, Mark. *The Making of the Chinese Industrial Workplace: State, Revolution, and Labor Management*. Cambridge, Eng.: Cambridge University Press, 2002.

Freedman, Maurice. *Chinese Lineage and Society: Fukien and Kwangtung*. London: Athlone Press, 1966.

————. *Lineage Organization in Southeastern China*. London: Athlone Press, 1958.

Fried, Morton. *The Fabric of Chinese Society: A Study of the Social Life of a Chinese County Seat*. London: Atlantic Press, 1956.

Friedman, Andrew. *Industry and Labor: Class Struggle at Work and Monopoly Capitalism*. London: Macmillan, 1977.

Friedman, Edward; Paul Pickowicz; and Mark Selden. *Chinese Village, Socialist State*. New Haven: Yale University Press, 1991.

————. *Revolution, Resistance and Reform in Village China*. New Haven: Yale University Press, 2005.

Gamble, Sydney. *North China Villages: Social, Political, and Economic Activities Before 1933*. Berkeley: University of California Press, 1963.

————. *Ting Hsien: A North China Rural Community*. New York: Institute of Pacific Relations, 1954.

Gan Cisen. *Zuijin 45 nianlai Sichuan sheng jinchukou maoyi tongji* (Import and export statistics of Sichuan during the last 45 years). Chongqing: Minsheng shiye gongsi jingji yanjiushi, 1936.

Gan Duan. *Jiajiang xian xiangtu zhilüe* (Abridged gazetteer of the localities of Jiajiang county). Jiajiang: Yitong shuju, 1948.

Gerth, Karl. *China Made: Consumer Culture and the Creation of the Nation*. Cambridge: Harvard University Asia Center, 2003.

Giersch, C. Pat. "A Motley Throng: Social Change on Southwest China's Early Modern Frontier, 1700–1880." *Journal of Asian Studies* 60, no. 10 (Feb. 2001): 67–94.

Graham, David Crockett. *Folk Religion in Southwest China*. Washington, DC: Smithsonian Institution, 1961.

Greenhalgh, Susan, and Edwin A. Winckler. *Governing China's Population: From Leninist to Neoliberal Biopolitics*. Stanford: Stanford University Press, 2005.

Gufosi stele. 1839. Transcribed stele text in the possession of the author.

Guojia tongjiju. *Zhongguo nongcun tongji nianjian, 1996 juan*. Beijing: Zhongguo tongji, 1997.

Hafter, Daryl M. "Women Who Wove in the Eighteenth-Century Silk Industry of Lyon." In *European Women and Preindustrial Craft*, ed. idem. Bloomington: Indiana University Press, 1995, 42–64.

Han Xiaorong. *Chinese Discourses on the Peasant, 1900–1949*. Albany: State University of New York Press, 2005.

Harrison, Henrietta. *A Man Awakened from Dreams: One's Man Life in a North China Village, 1857–1942*. Stanford: Stanford University Press, 2005.

————. "Village Industries and the Making of Rural-Urban Difference in Early Twentieth-Century Shanxi." In *How China Works: Perspectives on the Twentieth-Century Workplace*, ed. Jacob Eyferth. Milton Park, Eng.: Routledge, 2006, 25–40.

Hayford, Charles. *To the People: James Yen and Village China*. New York: Columbia University Press, 1990.

Henderson, Gail E.; Barbara Entwisle; Li Ying; Yang Mingliang; Xu Siyuan; and Zhai Fengying. "Re-drawing the Boundaries of Work: Views on the Meaning of Work (*Gongzuo*)." In *Re-drawing Boundaries: Work, House-hold, and Gender in China*, ed. Barbara Entwistle and Gail Henderson. Berkeley: University of California Press, 2000, 33–50.

Hershatter, Gail. *The Workers of Tianjin, 1900–1949*. Stanford: Stanford University Press, 1986.

Herzfeld, Michael. *The Body Impolitic: Artisans and Artifice in the Global Hierarchy of Value*. Chicago: University of Chicago Press, 2004.

Hinton, William. *Fanshen: A Documentary of Revolution in a Chinese Village*. New York: Vintage, 1966.

Honeyman, Katrina, and Jordan Goodman. "Women's Work, Gender Conflict, and Labour Markets in Europe, 1500–1900." In *Gender and History in Western Europe*, ed. Robert Shoemaker and Mary Vincent. London: Arnold, 1998, 353–76.

Honig, Emily. *Sisters and Strangers: Women in the Shanghai Cotton Mills, 1919–1949*. Stanford: Stanford University Press, 1986.

Hosie, Alexander. *Szechwan: Its Products, Industries, and Resources*. Shanghai: Kelly and Walsh, 1922.

Hsiang, C. Y. "Mountain Economy in Sichuan." *Pacific Affairs* 14, no. 4 (Dec. 1941): 448–62.

Hsu, Francis L. K. "Observations on Cross-Cousin Marriage in China." *American Anthropologist* 47 (Jan.–Mar. 1945): 83–103.

Hu Hsien-Chin. *The Common Descent Group in China and Its Functions*. New York: Wenner-Gren Foundation, 1948.

Hua Younian. "Jiajiang de zhiye yu jinrong" (The paper industry and finance in Jiajiang). *Sichuan jingji jikan* 1, no. 3 (1944): 415–19.

Huang Fuyuan. *Zhima haozi* (Bamboo work songs). Jiajiang: photocopied manuscript, n.d.

Huang, Philip C. C. *The Peasant Economy and Social Change in North China*. Stanford: Stanford University Press, 1985.

————. *The Peasant Family and Rural Development in the Yangzi Delta, 1350–1988*. Stanford: Stanford University Press, 1990.

Hung, Chang-Tai. *War and Popular Culture: Resistance in Modern China, 1937–1945*. Berkeley: University of California Press, 1994.

Hutchins, Edwin. "Learning to Navigate." In *Understanding Practice: Perspectives on Activity and Context*, ed. Seth Chaiklin and Jean Lave. Cambridge: Cambridge University Press, 1993, 35–63.

Ingold, Tim. *The Perception of the Environment: Essays on Livelihood, Dwelling, and Skill*. London: Routledge, 2000.

Jiadangqiao stele, Shiyan village, Macun township, Jiajiang. Stele text in possession of the author.

Jiajiang turang (Soil in Jiajiang). Jiajiang: Jiajiang xian nongyeju, 1984.

Jiajiang xian xiangtuzhi (Gazetteer of market towns in Jiajiang). Jiajiang, n.d. [late Qing].

Jiajiang xian 1994 nian nongye shengchan, nongcun jingji qingkuang (Agricultural production and the rural economic situation in Jiajiang county, 1994). Jiajiang: photocopied manuscript, 1995.

Jiajiang xian xiangzhen gaikuang (Basic facts on the townships and towns of Jiajiang county). Jiajiang: Jiajiang xian difangzhi bangongshi, 1991.

Jiang Huice. "Sichuan xi'nanqu jingji jianshe couyi" (My observations of economic reconstruction in the southwest of Sichuan). *Sichuan jingji jikan* 2, no. 3 (July 1945): 68–97.

Jing Jun. *The Temple of Memories: History, Power, and Morality in a Chinese Village*. Stanford: Stanford University Press, 1996.

Judd, Ellen. *Gender and Power in Rural North China*. Stanford: Stanford University Press, 1994.

Kane, Penny. *Famine in China, 1958–1961: Demographic and Social Implications*. Basingstoke, Eng.: Macmillan, 1988.

Kapp, Robert A. "Chungking as a Center of Warlord Power, 1926–1937." In *The Chinese City Between Two Worlds*, ed. Mark Elvin and G. William Skinner. Stanford: Stanford University Press, 1974, 143–70.

———. *Sichuan and the Chinese Republic, 1911–1938*. New Haven: Yale University Press, 1973.

Kipnis, Andrew B. *Producing Guanxi: Sentiment, Self, and Subculture in a North China Village*. Durham, NC: Duke University Press, 1997.

———. "Within and Against Peasantness: Backwardness and Filiality in Rural China." *Comparative Studies in Society and History*, no. 37 (1995): 110–35.

Kirby, William C. "Engineering China: Birth of the Developmental State, 1928–1937." In *Becoming Chinese: Passages to Modernity and Beyond*, ed. Wen-Hsin Yeh. Berkeley: University of California Press, 2000, 137–60.

————. *Germany and Republican China*. Stanford: Stanford University Press, 1984.

Knight, John, and Lina Song. *The Rural-Urban Divide: Economic Disparities and Interactions in China*. Oxford: Oxford University Press, 1999.

Koepp, Cynthia J. "The Alphabetical Order: Work in Diderot's *Encyclopédie*." In *Work in France: Representations, Meaning, Organization, and Practice*, ed. Steven L. Kaplan and Cynthia J. Koepp. Ithaca: Cornell University Press, 1986, 229–57.

Kotkin, Stephen. *Magnetic Mountain: Stalinism as a Civilization*. Berkeley: University of California Press, 1995.

Kraus, Richard Curt. *Brushes with Power: Modern Politics and the Chinese Art of Calligraphy*. Berkeley: University of California Press, 1991.

Kuhn, Philip. "Chinese Views of Social Stratification." In *Class and Social Stratification in Post-Revolutionary China*, ed. James L. Watson. Cambridge: Cambridge University Press, 1984, 16–28.

Landes, David S. "What Do Bosses Really Do?" *Journal of Economic History* 46, no. 3 (Sept. 1986): 585–623.

Lardy, Nicholas. *Agriculture in China's Modern Economic Development*. Cambridge: Cambridge University Press, 1983.

Lave, Jean, and Etienne Wenger. *Situated Learning: Legitimate Peripheral Participation*. Cambridge: University of Cambridge Press, 1991.

Leach, Edmund. *Pul Eliya: A Village in Ceylon*. Cambridge: Cambridge University Press, 1961.

Lean, Eugenia. "One Part Cow Fat, Two Parts Soda: Recipes for a New Urban Identity and the Gender of Science in 1910s China." Unpublished manuscript.

Leshan ribao. "Zhixiang daitouren—ji sheng laomo Shi Fuli" (Leader of the paper county: an interview with provincial labor model Shi Fuli). *Leshan ribao*, Oct. 9, 1995, 3.

Li Bozhong. *Jiangnan de zaoqi gongyehua* (Protoindustrialization in Jiangnan). Beijing: Shehui kexue wenxuan, 2000.

Li Jiwei. "Gailiang Jiajiang zaozhiye zhi wojian" (My opinions on reform of the Jiajiang paper industry). *Sichuan shanhou duban gongshu yuekan* 1, no. 1 (Sept. 1934): 17–20.

Li Liugen. "Fazhan tese jingji, peizhi zhizhu chanye" (Develop an economy with special characteristics, foster the main trade). In *Jiajiang nianjian* (Jiajiang yearbook) 1999, 138–39.

Li Shiping. *Sichuan renkoushi* (The demographic history of Sichuan). Chengdu: Sichuan daxue, 1987.

Li Yubing. "Jiajiang chanzhi ying shiyu heshi?" (When did papermaking in Jiajiang originate?). Jiajiang: photocopied manuscript, 1986.

Li Yubing and Lei Yinglan. "Zhiye jushang Xie Rongchang jianli" (A short life of the great paper trader Xie Rongchang). In *Jiajiang wenshi ziliao* (Materials on culture and history in Jiajiang). Jiajiang: Zhengxie wei-yuanhui, 1986, 1–9.

Liang Binwen. "Sichuan zhiye diaocha baogao" (Investigation report on Sichuan papermaking). *Jianshe tongxun* 1, no. 10 (1937): 15–30.

Little, Archibald John. *Mount Omi and Beyond: A Record of Travels on the Tibetan Border*. London: Heinemann, 1901.

Liu Hongkang. *Zhongguo renkou: Sichuan fence* (The population of China: volume on Sichuan). Beijing: Zhongguo caizheng jingji, 1988.

Liu, Hui-chen Wang. *The Traditional Chinese Clan Rules*. New York: J. J. Augustin, 1959.

Liu, Lydia H. *Translingual Practice. Literature, National Culture, and Translated Modernity: China, 1900–1937*. Stanford: Stanford University Press, 1995.

Liu Min. "Sichuan shehui jingji zhi lishi xingge yu gongye jianshe" (The historical character of Sichuan's society and economy and industrial reconstruction). *Sichuan jingji jikan* 2, no. 1 (Jan. 1945): 99–108.

Liu Shaoqi. "Guanyu shougongye hezuoshe wenti" (On the problems of handicraft cooperatives). In *Zhongguo shougongye hezuohua he chengzhen jiti gongye de fazhan* (The collectivization of China's handicrafts and the development of collective urban industries), ed. Zhonghua quanguo shougongye hezuo zongshe. Beijing: Dangshi, 1992, 1: 104–9.

———. "Guanyu xin Zhongguo de jingji jianshe fangzhen" (New China's economic development strategy). In *Zhongguo shougongye hezuohua he chengzhen jiti gongye de fazhan* (The collectivization of China's handicrafts and development of collective urban industries), ed. Zhonghua quanguo shougongye hezuo zongshe. Beijing: Dangshi, 1992, 1: 26–30.

Liu Shaoquan. *Jiajiang de zhiye yu guoji jiaoliu* (Jiajiang's paper trade and international exchange). Chengdu: Sichuan daxue, 1992.

Liu Ta-Chung and Kung-Chia Yeh. *The Economy of the Chinese Mainland: National Income and Economic Development, 1933–1959*. Princeton: Princeton University Press, 1965.

Liu, Tessa. *The Weaver's Knot: The Contradictions of Class Struggle and Family Solidarity in Western France, 1750–1914*. Ithaca, NY: Cornell University Press, 1994.

Liu Zidong. "Sanshisannian Jiajiang jingji dongtai" (Economic tendencies in Jiajiang in 1944). *Sichuan jingji jikan* 2, no. 2 (1945): 199–202.

Liu Zuoming. *Jiajiang xianzhi* (Jiajiang county gazetteer). 1935. Reprinted—Jiajiang: Jiajiang xian difangzhi bangongshi, 1985.

Lu Deheng. "Bian shougongzhi shengchan zhouqi 100 tian wei 3 tian" (Reduce the production cycle for handmade paper from 100 to 3 days). *Zaozhi gongye* 10 (1958): 19–20.

———. "Shougongzhi shengchan zhong de jishu gexin" (Technological innovation in handmade paper production). *Zaozhi gongye* 6 (1958): 12–13.

Lu Zijian. *Qingdai Sichuan caizheng shiliao* (Historical materials on taxation in Qing period Sichuan). 2 vols. Chengdu: Sichuan sheng shehui kexueyuan, 1988.

Lü Pingdeng. *Sichuan nongcun jingji* (Sichuan's rural economy). Shanghai: Shangwu, 1936.

Lüdtke, Alf. "What Happened to the 'Fiery Red Glow'? Workers' Experiences and German Fascism." In *History of Everyday Life: Reconstructing Historical Experiences and Ways of Life*, ed. idem. Princeton: Princeton University Press, 1995, 198–251.

Ma Mingzhang. "Luoshi quanshu, lin nong huanwei, shengtai buchang" (Clarify property rights, reverse the priority of forestry and agriculture, repair the environment). Unpublished conference paper, 1995.

Ma Youmei. "Jiajiang wupo wuli zhua jiyu quanxian gongye chanzhi caizheng shouru wenbu zengchang" (Jiajiang's Five Eradicate and Five Create, grasp opportunity, steadily increase industrial production value and the fiscal income of the entire county). In *Jiajiang nianjian* (Yearbook of Jiajiang), 2000, 153–54.

———. "Qiangji guben, wanshan fuwu, Jiajiang zhuoli suzao 'bei damen' xingxiang" (Strengthen the base, improve service, mobilize strength to erect a "Great North Gate" [to Leshan City] in Jiajiang). In *Jiajiang nianjian* (Yearbook of Jiajiang), 2000, 152–53.

MacGowan, D. J. "Chinese Guilds or Chambers of Commerce and Trade Unions." *Journal of the North China Branch of the Royal Asiatic Society*, no. 28 (1886): 133–92.

Macun xiang dishibao diaochapu (Survey chart of 10th *bao* of Macun township). N.d.

Maier, Charles S. "Between Taylorism and Technocracy: European Ideologies and the Vision of Industrial Productivity in the 1920s." *Journal of Contemporary History* 5, no. 2 (1970): 27–61.

Mann, Susan L. "Household Handicrafts and State Policy in Qing Times." In *To Achieve Security and Wealth: The Qing Imperial State and the Economy, 1644–1911*, ed. Jane Kate Leonard and John R. Watt. Ithaca: Cornell University Press, 1992, 75–95.

————. *Precious Records: Women in China's Long Eighteenth Century*. Stanford: Stanford University Press, 1997.

Mao Dun. *Spring Silkworms and Other Stories*. Beijing: Foreign Languages Press, 1979.

Mao Zedong. "How to Differentiate the Classes in the Rural Areas." Oct. 1933. In *Selected Works of Mao Tse-tung*. Beijing: Foreign Languages Press, 1967, 1: 137–39.

————. *Report from Xunwu*. Trans. and ed. Roger R. Thompson. Stanford: Stanford University Press, 1990.

————. "Speed up the Socialist Transformation of Handicrafts." Mar. 5, 1956. In *Selected Works of Mao Tse-tung*. Beijing: Foreign Languages Press, 1977, 5: 281–83.

Marglin, Frédérique Apffel, and Stephen A. Marglin. *Dominating Knowledge: Development, Culture, and Resistance*. Oxford: Clarendon Press, 1990.

Marglin, Stephen A. "Understanding Capitalism: Control Versus Efficiency." In *Power and Economic Institutions: Reinterpretations in Economic History*, ed. B. Gustafsson. Brookfield, VT: Elgar, 1991.

————. "What Do Bosses Do?" In *The Division of Labour: The Labour Process and Class-Struggle in Modern Capitalism*, ed. André Gorz. Brighton, Eng.: Harvester Press, 1974, 13–54.

Marshall, Thomas Humphrey. *Citizenship and Social Class, and Other Essays*. Cambridge: Cambridge University Press, 1950.

Martin, Michael. "Defining China's Rural Population." *China Quarterly*, no. 130 (1992): 392–401.

Marx, Karl. *Capital*, vol. 1. London: Dent, 1930.

————. "The Eighteenth Brumaire of Louis Bonaparte." In *The Marx-Engels Reader*, ed. Robert C. Tucker. New York: Norton, 1978, 594–617.

Marx, Karl, and Frederick Engels. "Manifesto of the Communist Party." In *The Marx-Engels Reader*, ed. Robert C. Tucker. New York: Norton, 1978, 469–500.

Masini, Federico. *The Formation of Modern Chinese Lexicon and Its Evolution Toward a National Language: The Period from 1840 to 1849*. Journal of Chinese Linguistics Monograph Series, 6. Berkeley: Project on Linguistic Analysis, University of California, 1993.

Mazumdar, Sucheta. *Sugar and Society in China: Peasants, Technology, and the World Market*. Cambridge: Harvard University Asia Center, 1998.

Medick, Hans. "Village Spinning Bees: Sexual Culture and Free Time Among Rural Youth in Early Modern Germany." In *Interest and Emotion: Essays on the Study of Family and Kinship*, ed. idem and David Warren

Sabean. Cambridge: Cambridge University Press; Paris: Editions de la Maison des Sciences de l'homme, 1984, 317–39.

Mendels, Franklin. "Proto-Industrialization: The First Phase of the Industrialization Process." *Journal of Economic History* 32, no. 1 (1972): 241–61.

Meskill, Johanna. *A Chinese Pioneer Family: The Lins of Wufeng, Taiwan, 1729–1895*. Princeton: Princeton University Press, 1979.

Mitchell, Timothy. *Rule of Experts: Egypt, Techno-Politics, Modernity*. Berkeley: University of California Press, 2002.

Mokyr, Joel. *The Gifts of Athena: Historical Origins of the Knowledge Economy*. Princeton: Princeton University Press, 2002.

More, Charles. *Skill and the English Working Class, 1870–1914*. London: Croom Helm, 1980.

Morgan, Stephen L. "Scientific Management in China, 1910–1930s." [University of Melbourne] Department of Management Working Paper Series, Working Paper 2003/10012, http://www.management.unimelb .edu.au/staff/paper/Morgan%20manuscript.pdf, accessed Feb. 4, 2007.

Morris-Suzuki, Tessa. *The Technological Transformation of Japan: From the Seventeenth to the Twenty-First Century*. Cambridge: Cambridge University Press, 1994.

Mueggler, Erik. *The Age of Wild Ghosts: Memory, Violence, and Place in Southwest China*. Berkeley: University of California Press, 2001.

———. "The Poetics of Grief and the Price of Hemp in Southwest China." *Journal of Asian Studies* 57, no. 4 (Nov. 1998): 979–1008.

Naquin, Susan. *Peking: Temples and City Life, 1400–1900*. Berkeley: University of California Press, 2001.

Nisbet, Robert. "Citizenship: Two Traditions." In *Citizenship: Critical Concepts*, ed. Bryan S. Turner and Peter Hamilton. London: Routledge, 1994, 1: 7–23.

Oi, Jean C. *State and Peasant in Contemporary China: The Political Economy of Village Government*. Berkeley: University of California Press, 1989.

Pan Jixing. *Zhongguo zaozhi jishu shigao* (A draft history of Chinese papermaking technology). Beijing: Wenwu, 1975.

Pasternak, Burton. *Kinship and Community in Two Chinese Villages*. Stanford: Stanford University Press, 1972.

Peng Xizhe. *Demographic Transition in China: Fertility Trends Since 1954*. Oxford: Clarendon Press, 1991.

Peng Zeyi. *Zhongguo jindai shougongye shi ziliao* (Materials on the history of handicrafts in modern China). 4 vols. Beijing: Sanlian, 1957.

Perry, Elizabeth J. "From Native Place to Workplace: Labor Origins and Outcomes of China's *Danwei* System." In *Danwei: The Changing Chinese*

Workplace in Historical and Comparative Perspective, ed. idem and Lü Xiaobo. Armonk, New York: M. E. Sharpe, 1997, 42–59.

———. "Introduction: Putting Class in Its Place: Bases of Worker Identity in East Asia." In *Putting Class in Its Place: Worker Identity in East Asia*, ed. idem. Berkeley: University of California Press, 1996, 1–10.

———. *Shanghai on Strike: The Politics of Chinese Labor*. Stanford: Stanford University Press, 1993.

Pieke, Frank N. "The Genealogical Mentality in Modern China." *Journal of Asian Studies* 62, no. 1 (Jan. 2003): 101–28.

Piore, Michael J., and Charles Sabel. *The Second Industrial Divide: Possibilities for Prosperity*. New York: Basic Books, 1984.

Pomeranz, Kenneth. *The Great Divergence: Europe, China, and the Making of the Modern World Economy*. Princeton: Princeton University Press, 2000.

———. "Women's Work, Family, and Economic Development in Europe and East Asia." In *The Resurgence of East Asia: 500, 150 and 50 Year Perspectives*, ed. Giovanni Arrighi, Takeshi Hamashita, and Mark Selden. London: Routledge, 2003, 124–72.

Portelli, Alessandro. *The Death of Luigi Trastulli and Other Stories: Form and Meaning in Oral History*. Albany: State University of New York Press, 1991.

Potter, Sulamith Heins. "The Position of Peasants in Modern China's Social Order." *Modern China* 9, no. 4 (Oct. 1983): 465–99.

Potter, Sulamith Heins, and Jack M. Potter. *China's Peasants: The Anthropology of a Revolution*. Cambridge: Cambridge University Press, 1990.

Pun Ngai. "Becoming *Dagongmei* (Working Girls): The Politics of Identity and Difference in Reform China." *China Journal* 42 (July 1999): 1–18.

———. *Made in China: Subject, Power, and Resistance in a Global Workplace*. Durham, NC: Duke University Press, 2005.

Qiu Xian. "Zhenxing zaozhi gongye yu shougong zaozhi zhi gaijin" (Reviving the paper industry and the reform of handicraft papermaking). *Xinan shiye tongxun* 3, no. 1 (Jan. 1942): 19–20.

Quan Hansheng. *Zhongguo hanghui zhidu shi* (A history of China's guild system). Shanghai: Shihuo, 1935.

Rabinbach, Anson. *The Human Motor: Energy, Fatigue, and the Origins of Modernity*. Berkeley: University of California Press, 1992.

Rawski, Evelyn S. "Agricultural Development in the Han River Highlands." *Ch'ing-shih wen-t'i* 3, no. 4 (1975): 63–81.

Ren Zhijun. "Jiajiang shougong zhi de chanxiao gaikuang" (Basic facts on the production and marketing of Jiajiang handicraft paper). In *Jiajiang wenshi ziliao* (Materials on culture and history in Jiajiang). Jiajiang: Zhengxie weiyuanhui, 1986, 1–9.

———. "Yishu Jiajiang yi zaozhi wei zhongxin de jingji shilüe" (Remembering the paper-centered economy of Jiajiang). Jiajiang: photocopied manuscript, n.d.

Riskin, Carl. "China's Rural Industries: Self-reliant Systems or Independent Kingdoms?" *China Quarterly*, no. 73 (1978): 77–98.

———. "Small Industry and the Chinese Model of Development." *China Quarterly*, no. 46 (1971): 245–73.

Rofel, Lisa. *Other Modernities: Gendered Yearnings in China After Socialism.* Berkeley: University of California Press, 1999.

Rogger, Hans. "Amerikanizm and the Economic Development of Russia." *Comparative Studies in Society and History* 23, no. 3 (July 1981): 382–420.

Rowe, William T. *Crimson Rain: Seven Centuries of Violence in a Chinese County.* Stanford: Stanford University Press, 2007.

———. *Hankow: Conflict and Community in a Chinese City, 1769–1895.* Stanford: Stanford University Press, 1989.

———. *Saving the World: Chen Hongmou and Elite Consciousness in Eighteenth-Century China.* Stanford: Stanford University Press, 2001.

Ruf, Gregory A. *Cadres and Kin: Making a Socialist Village in West China, 1921–1991.* Stanford: Stanford University Press, 1998.

Rule, John. "The Property of Skill in the Period of Manufacture." In *The Historical Meaning of Work*, ed. Patrick Joyce. Cambridge: Cambridge University Press, 1987, 98–118.

Sabel, Charles. *Work and Politics: The Division of Labor in Industry.* Cambridge: Cambridge University Press, 1982.

Sabel, Charles F., and Jonathan Zeitlin. "Historical Alternatives to Mass Production: Politics, Markets, and Technology in Nineteenth-Century Industrialization." *Past and Present* 108 (Aug. 1985): 133–76.

Sabel, Charles F., and Jonathan Zeitlin, eds. *Worlds of Possibilities: Flexibility and Mass Production in Western Industrialization.* Cambridge: Cambridge University Press; Paris: Musée de l'homme, 1997.

Santos, Gonçalo Duro dos. "The Anthropology of Chinese Kinship: A Critical Overview." *European Journal of East Asian Studies* 5, no. 2 (2006): 275–333.

Schran, Peter. "Handicrafts in Communist China." *China Quarterly*, no. 17 (Jan.–Mar. 1964): 151–73.

Schwartz, Benjamin. *In Search of Wealth and Power: Yen Fu and the West.* Cambridge: Harvard University Press, Belknap Press, 1964.

Scott, James C. *Seeing Like a State.* New Haven: Yale University Press, 1998.

Scott, Joan W. "L'ouvrière: mot impie et sordide." In Joan W. Scott, *Gender and the Politics of History*. New York: Columbia University Press, 1999, 113–63.

Selden, Mark. *The People's Republic of China: A Documentary History of Revolutionary Change*. New York: Monthly Review Press, 1980.

———. *The Political Economy of Chinese Development*. Armonk, NY: M. E. Sharpe, 1993.

Sewell, William H. *Work and Revolution in France: The Language of Labor from the Old Regime to 1848*. Cambridge: Cambridge University Press, 1980.

Sheng Yi and Yuan Dingji. "Jiajiang zaozhi" (Jiajiang papermaking). *Hansheng*, no. 77 (1995): 1–43. Republished as Wu Meiyun. *Jiajiang zaozhi*. Taibei: Hansheng zazhishe, 1995.

Shiga, Shuzo. "Family Property and the Law of Inheritance in Traditional China." In *Chinese Family Land and Social Change in Historical and Comparative Perspective*, ed. David C. Buxbaum. Seattle: University of Washington Press, 1978, 109–50.

Sichuan sheng Jiajiang xian bianshi xiuzhi weiyuanhui, ed. *Jiajiang xian nianjian* (Yearbook of Jiajiang county), 1987–2006.

Sichuan sheng shougongzhi shengchan jishu jingyan jiaoliuhui. "Shougongzhi shengchan zhong shiyong daiyong yuanliao he gaijin shengchan gongju de jingyan" (Experiences on the use and replacement of raw materials in handmade paper production and the improvement of production tools). *Zaozhi gongye*, no. 4 (1958): 14–15.

Sichuan sheng zhengfu. Jiansheting. *Sichuan sheng E-Jia-Le sanxian chaye diaocha baogao* (Survey report on tea production in Emei, Jiajiang, and Leshan). Chengdu: Jianshe congshu, 1939.

Sichuan xinwen chubanju. Shizhi bianzou weiyuanhui. *Sichuan xinwen chuban shiliao* (Materials on the history of the press and publishing in Sichuan). Chengdu: Sichuan renmin, 1976.

Siegelbaum, Lewis, and Ronald G. Suny, ed. *Making Workers Soviet*. Ithaca, NY: Cornell University Press, 1994.

Sigaut, François. "Technology." In *Companion Encyclopedia of Anthropology: Humanity, Culture, and Social Life*, ed. Tim Ingold. London: Routledge, 1994, 420–59.

Sigurdson, Jon. *Rural Industrialization in China*. Cambridge: Council on East Asian Studies, Harvard University, 1977.

Skinner, G. William. "Aftermath of Communist Liberation in the Chengdu Plain." *Pacific Affairs* 24, no. 1 (1951): 61–76.

——. "Cities and the Hierarchy of Local Systems." In *The City in Late Imperial China*, ed. idem. Stanford: Stanford University Press, 1977, 275–351.

——. "Marketing and Social Structure in Rural China." 3 pts. *Journal of Asian Studies* 24, no. 1 (1964): 5–43; 24, no. 2 (1964): 195–228; 25, no. 1 (1965): 363–99.

——. "Regional Urbanization in Nineteenth-Century China." In *The City in Late Imperial China*, ed. idem. Stanford: Stanford University Press, 1977, 211–49.

——. "Sichuan's Population in the Nineteenth Century: Lessons from Disaggregated Data." *Late Imperial China* 8, no. 1 (1987): 1–79.

Solinger, Dorothy J. *Contesting Citizenship in Urban China: Peasant Migrants, the State, and the Logic of the Market.* Berkeley: University of California Press, 1999.

——. *Regional Government and Political Integration in Southwest China, 1949–1954.* Berkeley: University of California Press, 1977.

Somers, Margaret R. "The 'Misteries' of Property: Relationality, Rural Industrialization, and Community in Chartist Narratives of Political Rights." In *Early Modern Conceptions of Property*, ed. John Brewer and Susan Staves. London: Routledge, 1996, 63–92.

——. "Rights, Relationality, and Membership: Rethinking the Making and Meaning of Citizenship." *Law and Social Inquiry* 19, no. 1 (Winter 1994): 63–112.

Sonenscher, Michael. *The Hatters of Eighteenth-Century France.* Berkeley: University of California Press, 1987.

Stapleton, Kristin. "Urban Politics in an Age of 'Secret Societies': The Cases of Shanghai and Chengdu." *Republican China* 22, no. 1 (1996): 23–64.

Stone, Glenn Davis. "Agricultural Deskilling and the Spread of Genetically Modified Cotton in Warangal." *Current Anthropology* 48, no. 1 (Feb. 2007): 67–103.

Su Shiliang. "Jiajiang xian zhiye zhi gaikuang" (The state of papermaking in Jiajiang county). *Nongye zazhi* 1, no. 1 (1923): *diaocha* section, 7–19.

Thompson, Paul. *The Nature of Work: An Introduction to Debates on the Labour Process.* London: Macmillan, 1983.

Ts'ien, Tsuen-hsuin. *Paper and Printing.* Vol. 5, pt. 1, of *Science and Civilisation in China*, ed. Joseph Needham. Cambridge: Cambridge University Press, 1985.

Tu Song. *Jiajiang xianzhi* (Gazetteer of Jiajiang). 1813.

Veilleux, Louis. *The Paper Industry in China from 1949 to the Cultural Revolution*. Toronto: University of Toronto, 1978.

Vermeer, E. B. "The Mountain Frontier in Late Imperial China: Economic and Social Developments in the Dabashan." *T'oung Pao* 77, no. 4–5 (1991): 301–35.

Walker, Kenneth R. *Food Grain Procurement and Consumption in China*. Cambridge: Cambridge University Press, 1984.

Wang Di. "Qingdai Sichuan renkou, gengdi ji liangshi wenti" (The question of population, farmland, and grain in Qing-period Sichuan). 2 pts. *Sichuan daxue xuebao*, no. 3 (1989): 90–105; no. 4 (1989): 73–87.

Wang Fei-ling. "Reformed Migration Control and New Targeted People: China's Hukou Sytem in the 2000s." *China Quarterly*, no. 177 (2004): 115–32.

Wang Gang. "Qingdai Sichuan de zaozhi yu chuban yinshua" (Papermaking and printing in Qing-period Sichuan). In *Qingdai Sichuan shi* (History of Sichuan in the Qing period), ed. idem. Chengdu: Sichuan renmin, 1991, 688–707.

Wang Haibo. *Xin Zhongguo gongye jingji shi, 1949:10–1957* (History of the industrial economy of New China, October 1949 to 1957). Beijing: Jingji guanli chubanshi, 1994.

Wang Huning. *Dangdai Zhongguo cunluo jiazu wenhua* (Village and lineage culture in contemporary China). Shanghai: Shanghai renmin, 1991.

Wang Lixian. *Sichuan gonglu jiaotong shi* (History of roads and transport in Sichuan). Chengdu: Sichuan renmin, 1989.

Wang Shaoguang. "The Politics of Private Time: Changing Leisure Patterns in Urban China." In *Urban Spaces in Contemporary China: The Potential for Autonomy and Community in Post-Mao China*, ed. Deborah Davis, Richard Kraus, Barry Naughton, and Elizabeth Perry. Washington, DC: Woodrow Wilson Center Press and Cambridge University Press, 1995, 149–72.

Wang Shaoquan. *Sichuan neihe hangyun shi* (History of river transportation in Sichuan). Chengdu: Sichuan renmin, 1989.

Wang Shugong, ed. *Jiajiang xianzhi* (Gazetteer of Jiajiang county). Chengdu: Sichuan renmin, 1989.

Wang Xiaoqiang and Bai Nanfeng. *The Poverty of Plenty*. Basingstoke, Eng.: Macmillan, 1991.

Watson, James L. "Chinese Kinship Reconsidered: Anthropological Perspectives on Historical Research." *China Quarterly*, no. 92 (Dec. 1982): 589–622.

Watson, Rubie S. *Inequality Between Brothers: Class and Kinship in South China.* Cambridge: Cambridge University Press, 1985.

―――. "The Named and the Nameless: Gender and Person in Chinese Society." *American Ethnologist* 13, no. 4 (1986): 619–31.

Weber, Eugen. *Peasants into Frenchmen: The Modernization of Rural France, 1870–1914.* Stanford: Stanford University Press, 1976.

Weber, Max. *The Religion of China: Confucianism and Taoism.* New York: Free Press, 1968.

Willis, Paul. *Learning to Labour: How Working Class Kids Get Working Class Jobs.* Aldershot, Eng.: Gower, 1977.

Wolf, Arthur P., and Chieh-shan Huang. *Marriage and Adoption in China, 1845–1945.* Stanford: Stanford University Press, 1980.

Wong, Christine P. W. "Central-Local Relations in an Era of Fiscal Decline: The Paradox of Fiscal Decentralization in Post-Mao China." *China Quarterly,* no. 128 (1991): 691–715.

Wong, R. Bin. *China Transformed: Historical Change and the Limits of European Experience.* Ithaca, NY: Cornell University Press, 1997.

Wood, Stephen, ed. *The Degradation of Work: Skill, Deskilling, and the Labour Process.* London: Hutchinson, 1982.

Wright, Tim. "Distant Thunder: The Regional Economies of Southwest China and the Impact of the Great Depression." *Modern Asian Studies* 34, no. 3 (July 2000): 697–738.

―――. "The Spiritual Heritage of Chinese Capitalism: Recent Trends in the Historiography of Chinese Enterprise Management." *Australian Journal of Chinese Affairs,* no. 19/20 (1988): 185–214.

Wu Meiyun. *Jiajiang zaozhi* (Jiajiang papermaking). Taibei: Hansheng zazhishe, 1995. Originally published as Sheng Yi and Yuan Dingji. "Jiajiang zaozhi." *Hansheng,* no. 77 (1995): 1–43.

Xie Changfu. "Jiajiang xian Huatou xiangzhi" (Gazetteer of Huatou township, Jiajiang county). Jiajiang: photocopied manuscript, 1988.

Yan Hairong. "Neoliberal Govermentality and Neohumanism: Organizing *Suzhi*/Value Flow Through Labor Recruitment Networks." *Cultural Anthropology* 18, no. 4 (2003): 493–523.

Yan Yunxiang. *Private Life Under Socialism: Love, Intimacy, and Family Change in a Chinese Village, 1949–1999.* Stanford: Stanford University Press, 2003.

Yang Bingwen. "Zhuming da caohu Shi Ziqing jianli" (A short life of the famous papermaker Shi Ziqing). In *Jiajiang wenshi ziliao* (Materials on culture and history in Jiajiang). Jiajiang: Zhengxie weiyuanhui, 1986, 26–27.

Yang Dali. *Calamity and Reform in China: State, Rural Society, and Institutional Change Since the Great Leap Famine.* Stanford: Stanford University Press, 1996.

Yang, Martin. *A Chinese Village: Taitou, Shandong Province.* New York: Columbia University Press, 1945.

Yang Minchuan. "Reshaping Peasant Culture and Community: Rural Industrialization in a Chinese Village." *Modern China* 20, no. 2 (1994): 157–79.

Yiwanshui stele. 1855. Stele text copied by the Jiajiang Culture Bureau; original stele no longer exists.

Yuan Peng. "Capital Formation in Rural Enterprises." In *Rural Enterprises in China*, ed. Christopher Findlay, Andrew Watson, and Harry X. Wu. New York: St. Martin's Press, 1994, 93–116.

Zanasi, Margherita. "Exporting Development: The League of Nations and Republican China." *Comparative Studies in Society and History* 49, no. 1 (2007): 143–69.

———. *Saving the Nation: Economic Modernity in Republican China.* Chicago: University of Chicago Press, 2006.

Zelin, Madeleine. *The Merchants of Zigong: Industrial Entrepreneurship in Early Modern China.* New York: Columbia University Press, 2005.

Zhai Shiyuan. "Wo suo zhidao de Jiajiang zhi zai Kunming xiaoshou de gaikuang" (Conditions of sales of Jiajiang paper in Kunming as I knew them). Jiajiang: photocopied manuscript, n.d.

Zhang, L., and Simon X. B. Zhao. "Re-examining China's 'Urban' Concept and the Level of Urbanization." *China Quarterly*, no. 154 (1998): 330–81.

Zhang Li. *Strangers in the City: Reconfigurations of Space, Power, and Social Networks Within China's Floating Population.* Stanford: Stanford University Press, 2001.

Zhang Ning. *Tudi de huanghun* (Dusk of the Earth). Beijing: Dongfang, 2005.

Zhang Wanshu. "Guanyu zhenxing Jiajiang zaozhide jianyi" (A proposal to revive Jiajiang papermaking). *Weiyuan zhi sheng*, no. 3 (July 25, 1991): 1–7.

———. "Qingdai Jiajiang zaozhi chutan" (Observations on papermaking in Qing-period Jiajiang). In *Qingdai de bianjiang kaifa* (The opening of the borderlands in the Qing period), ed. Wang Rongsheng and Wang Gang. Chengdu: Sichuan shehui kexueyuan, 1991.

Zhang Wenhua. "Jiajiang xian Macun xiangzhi" (Gazetteer of Macun township, Jiajiang county). Jiajiang: photocopied manuscript, 1990.

Zhang Xiaomei. *Sichuan jingji cankao ziliao* (Reference material on the economy of Sichuan). Chongqing: Zhongguo guomin jingji yanjiusuo, 1935.

Zhang Xuejun and Zhang Lihong. *Sichuan jindai gongyeshi* (History of Sichuan's early modern industry). Chengdu: Sichuan renmin, 1990.

Zhong Chongmin, Zhu Shouren, and Li Quan. *Sichuan shougong zhiye diaocha baogao* (Report on the handicraft paper industry of Sichuan). Chongqing: Zhongguo nongmin yinhang jingji yanjiusuo, 1943.

Zhong[guo] gong[chandang]. Zhongyang [weiyuanhui]. "Zhonggong Zhongyang guanyu chengxiang shougongye ruogan zhengce wenti de guiding—shixing cao'an" (Regulations of the CCP Central Committee on some policy questions regarding urban and rural handicrafts—provisional draft). In *Zhongguo shougongye hezuohua he chengzhen jiti gongye de fazhan* (The collectivization of China's handicrafts and the development of urban collective industries), ed. Zhonghua quanguo shougongye hezuo zongshe. Beijing: Dangshi, 1992, 2: 245–55.

——. "Zhonggong Zhongyang guanyu xunsu huifu he jinyibu fazhan shougongye shengchan de zhishi" (Directive of the CCP Central Committee to Quickly Restore and Further Develop Handicraft Production, August 1959). In *Zhongguo shougongye hezuohua he chengzhen jiti gongye de fazhan* (The collectivization of China's handicrafts and the development of urban collective industries), ed. Zhonghua quanguo shougongye hezuo zongshe. Beijing: Dangshi, 1992, 2:184–94.

Zhongguo renmin yinhang. Jiajiang zhihang. "Jiajiang zhiye shengchan jingying de diaocha baogao" (Survey report on production and management of the Jiajiang paper industry). *Diaoyan yu xinxi*, no. 39. Jiajiang: photocopied manuscript, Aug. 28, 1998.

Zhonghua quanguo shougongye hezuo zongshe. "Quanguo shougongye hezuo zongshe guanyu zhengdun, gonggu, tigao shougong hezuoshe de zhishi" (Directive by the All-China Handicraft Cooperative regarding the rectification, consolidation, and improvement of handicraft cooperatives). In *Zhongguo shougongye hezuohua he chengzhen jiti gongye de fazhan* (The collectivization of China's handicrafts and the development of urban collective industries), ed. Zhonghua quanguo shougongye hezuo zongshe. Beijing: Dangshi, 1992, 2: 291–96.

Zhonghua quanguo shougongye hezuo zongshe and Zhonggong Zhongyang dangshi yanjiushi, eds. *Zhongguo shougongye hezuohua he chengzhen jiti de fazhan* (The collectivization of China's handicrafts and the development of urban collective industries). 3 vols. Beijing: Dangshi, 1992–94.

Zhongyang renmin zhengfu. "Zhonghua renmin gongheguo tudi gaige fa" (Land reform law of the People's Republic of China). In *Tudi gaige zhongyao wenxian huiji* (Major documents on land reform). Beijing: Renmin, 1950, 2–10.

———. "Zhongyang renmin zhengfu Zhengwuyuan guanyu huafen nongcun jieji chengfen de jueding" (Decision of the Government Administration Council of the Central People's Government on how to differentiate rural classes). In *Tudi gaige zhongyao wenxian huiji* (Major documents on land reform). Beijing: Renmin, 1950, 33–59.

Zhongyang shougongye guanli zongju. Quanguo shougongye hezuo zongshe. "Guanyu 1963 nian jinyibu kaizhan zhengshe he zengchan jieyue yundong de zhishi" (Directive by the Central Handicraft Management Bureau and the All-China Handicraft Cooperative regarding the further extension of the movement for rectifying cooperatives, increasing production, and economizing). In *Zhongguo shougongye hezuohua he chengzhen jiti gongye de fazhan* (The collectivization of China's handicrafts and the development of collective urban industries), ed. Zhonghua quanguo shougongye hezuo zongshe. Beijing: Dangshi, 1992, 2: 307–11.

Zhou Kaiqing. *Sichuan jingji zhi* (Economic gazetteer of Sichuan). Taibei: Taiwan Shangwu, 1972.

Zhu De. "Ba shougongyezhe zuzhi qilai, zou shehuizhuyi daolu" (Organize artisans, take the road of socialism). In *Zhongguo shougongye hezuohua he chengzhen jiti gongye de fazhan* (The collectivization of China's handicrafts and the development of collective urban industries), ed. Zhonghua quanguo shougongye hezuo zongshe. Beijing: Dangshi, 1992, 1: 100–103.

Index

Acetylene factory (Shiyan), 192–97 *passim*, 278–79n23

Adoption, *see* Agnatic adoption

Advanced production cooperatives (APCs), 129, 130, 132

Advance sales (*yuhuo*), 83–84, 85–86, 111

Agnatic adoption, 31–32, 64, 174, 215

Agnatic kinship groups: *beifen* (distinction between generations), 58, 63–65, 66–67, 208, 214–16, 230–31; declining solidarity, 205; marriage restrictions, 213; papermakers, 7, 53; skill transmission within, 6–7, 41–42, 66, 99, 209–11; women's position, 65, 230

Agriculture: advanced production cooperatives, 129; cash crops, 93, 147, 156, 229; collectivization, 129; cooperatives, 113, 118; Great Leap Forward policies, 140–41; irrigation, 140–41, 200; local self-sufficiency policies, 147; state policies, 93. *See also* Grain; Rice

Ancestors: graves, 61, 62; offerings to, 4, 62; renewed interest in, 208; worship, 62, 65

Ancestral halls: construction, 217; in Jiajiang, 48–49, 60, 61–62; in Mianzhupu, 61–62, 211; rebuilt, 208

Ancestral tablets, 62

Anhui paper (*xuanzhi*), 76, 165–66, 171

APCs, *see* Advanced production cooperatives

Apprenticeships, 36, 163, 220

Aristotle, 7

Artisans: assigned to agricultural cooperatives, 118; Communist classifications, 124, 150; cooperatives, 118; guilds, 94, 220, 222; official registration, 118; rural, 8; social position, 118–19. *See also* Craft industries

Artists: demand for paper, 75–76; papermakers and, 75, 105. *See also* Calligraphers

Art paper, 158, 159

Auto-organization, 220–21, 223

Babbage, Charles, 11

Baimapu village, 51, 62

Bamboo: cutting, 25; dried, 165, 166, 170; fiber yield for

Harvard East Asian Monographs
(*out-of-print)